PLASTICS

The MIT Press Essential Knowledge Series

A complete list of books in this series can be found online at https://mitpress.mit.edu/books/series/mit-press-essential-knowledge-series.

PLASTICS

IMARI WALKER-FRANKLIN
AND JENNA JAMBECK

The MIT Press | Cambridge, Massachusetts | London, England

The MIT Press would like to thank the anonymous peer reviewers who provided comments on drafts of this book. The generous work of academic experts is essential for establishing the authority and quality of our publications. We acknowledge with gratitude the contributions of these otherwise uncredited readers.

This book was set in Chaparral Pro by Westchester Publishing Services. Printed and bound in the United States of America.

Library of Congress Cataloging-in-Publication Data

Names: Walker-Franklin, Imari, author. | Jambeck, Jenna, author.
Title: Plastics / Imari Walker-Franklin and Jenna Jambeck.
Description: Cambridge, Massachusetts : The MIT Press, [2023] | Series: The MIT press essential knowledge series | Includes bibliographical references and index.
Identifiers: LCCN 2022055535 (print) | LCCN 2022055536 (ebook) | ISBN 9780262547017 | ISBN 9780262377065 (epub) | ISBN 9780262377058 (pdf)
Subjects: LCSH: Plastics—Environmental aspects. | Plastic scrap—Environmental aspects. | Substitute products.
Classification: LCC TD195.P52 W35 2023 (print) | LCC TD195.P52 (ebook) | DDC 620.1/9230286—dc23/eng/20221129
LC record available at https://lccn.loc.gov/2022055535
LC ebook record available at https://lccn.loc.gov/2022055536

10 9 8 7 6 5 4 3 2 1

publication supported by a grant from
The Community Foundation for Greater New Haven
as part of the **Urban Haven Project**

CONTENTS

SERIES FOREWORD

The MIT Press Essential Knowledge series offers accessible, concise, beautifully produced pocket-size books on topics of current interest. Written by leading thinkers, the books in this series deliver expert overviews of subjects that range from the cultural and the historical to the scientific and the technical.

In today's era of instant information gratification, we have ready access to opinions, rationalizations, and superficial descriptions. Much harder to come by is the foundational knowledge that informs a principled understanding of the world. Essential Knowledge books fill that need. Synthesizing specialized subject matter for nonspecialists and engaging critical topics through fundamentals, each of these compact volumes offers readers a point of access to complex ideas.

1

INTRODUCTION

Wherever you are right now, look around. Do you perhaps have the book in one hand and your phone made of various metals, plastics, and glass in the other? Are you possibly perched on a park bench made of recycled plastic bags or on your polyurethane mattress? Did you recently flip on a light switch to read this? We come into contact daily with many items made of plastics—such as your phone (or parts of your phone), carpet, furniture, and light switches. Some plastic items might not be obvious including multi-materials containing plastic. For example, your clothing might be a polyester blend or fleece, a disposable coffee cup is paper lined with polyethylene, and metallic-looking snack bags are plastics with a very thin layer of aluminum.

While "plastic" is still used as an adjective in soil science, it is now a noun that is most known as "any of numerous organic synthetic or processed materials that are mostly

Before the word plastic became what we know it as today, it was used as an adjective to describe something "easily shaped or molded."

thermoplastic or thermosetting polymers of high molecular weight and that can be made into objects, films, or filaments."[1] According to Google, the use of the word plastic correlates with the timeline of the history and development of the common material we use today (figure 1). Since it was discovered in the 1860s and then brought to the public marketplace in the 1950s in numerous forms, plastic has become ubiquitous in our everyday lives.

Plastic, as the name of the material we now know, literally exploded onto the marketplace when humans were looking for a replacement for natural materials like ivory. The feedstock for most plastics today is fossil fuels, oil, and gas, but the invention of plastics came from a raw plant material, cellulose. Cellulose is the most prevalent organic compound on earth, as it is found in wood and cotton. It is composed of a chain of sugar molecules and makes up green plant cell walls. Paper derived from wood is made flexible by cellulose. Other natural materials that can be molded or shaped are amber and rubber (made from tree sap).

A German chemistry professor, Christian Friedrich Schönbein, found that the cellulose in cotton, when mixed with sulfur and nitric acid, became highly flammable and was termed "guncotton" or nitrocellulose. Partially nitrated cellulose was a bit more stable, termed pyroxylin, and was the basis for two plastic materials invented in the early 1860s—Parkesine and Xylonite—neither of which remained in the marketplace long. However, when

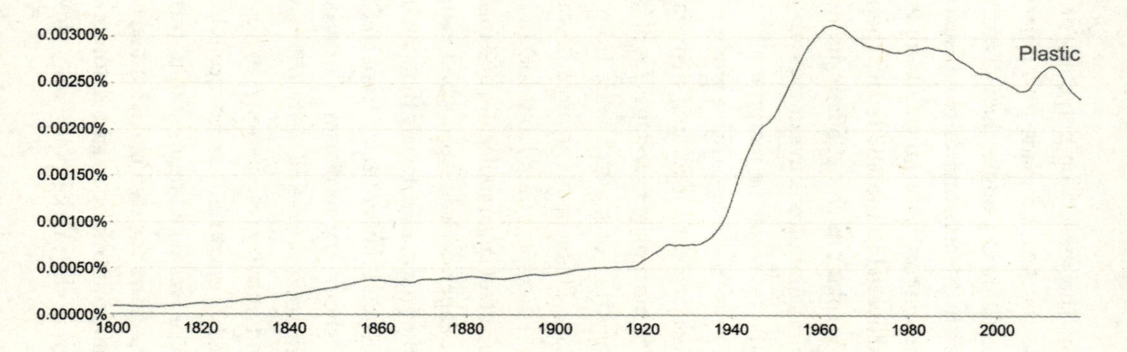

Figure 1 The use of the word "plastic" over time, according to Google Books.

dissolved in organic solvents, nitrocellulose became a very thick liquid termed collodion, which was also flammable and not very useful yet. The addition of camphor to the mixture resulted in a more stable moldable plastic-like material. John Wesley Hyatt, an American inventor, named this mixture "celluloid" and submitted it as an entry to win a $10,000 competition for finding a replacement for ivory in billiard balls. John won the competition, even though he never collected the prize money. These materials began to rapidly replace ivory. And while celluloid, popular in billiard balls, combs, and more, was more stable than nitrocellulose, the celluloid billiard balls could still be explosive while in use. Additionally, nitrocellulose was a biobased polymer originating from cellulose plant material; full synthetic oil and gas-sourced plastic materials came later.

In the early 1900s, like several other scientists, Leo Baekeland was developing new materials using phenol and formaldehyde mixtures. During this process, he added various temperatures and pressures. In 1909, he publicly announced his invention Bakelite, the world's first synthetic plastic material. Leo describes the polymerization process as akin to biological processes that take after the "phenolic nature of resinous substances" found in plants like Japanese Lacquer. In what is now an ironic twist, the process of making the first fully synthetic plastic material was, in his words, mimicking nature through biomimicry.

Bakelite had advantages over celluloid in that, once molded, the product maintained its shape when exposed to heat or chemicals. Since this thermoset plastic was heat tolerant, chemical resistant, and nonconductive, Bakelite was quickly implemented in many applications in the electrical and automotive industry, technological infrastructure (telephones, washing machine parts, etc.), and novelty items (jewelry, cosmetic containers, etc.). However, Bakelite could not be colored easily, and in some cases, sawdust was a filler used to make it opaque. And, as a thermoset, it could not be melted and remolded again. Eventually, thermoplastic materials that could be molded and remelted, like polystyrene and nylon, were invented. And these plastic materials had better aesthetics and became more widely used.

Plastics, as a new material, were quickly taken up for use by the military, especially in World War II when parachutes, helmet parts, gun pieces and housings, aircraft components, and even the combs issued to soldiers were all made of new plastic materials.[2] After the war, companies like DuPont that were making plastic materials were eager to see the products adopted by civilian society as well. And they got their wish through marketing a new disposable society to the public, as the infamous cover of *Life Magazine* showed in August 1955: "Throwaway Living." There was no denying the usefulness of plastic materials (moldable, durable, and light), and they expanded quickly to household goods, initially replacing those natural materials like ivory and rubber. A great example is cellophane that was invented in the early 1900s and used to store food.

The packaging sector has been the most significant sector of plastics use since 1950. Several factors drive the use in this sector. First, the world has become increasingly globalized. Since the quantity of goods and food that are transported globally is significant, the packaging sector has especially embraced plastic as a functional material. It is durable to protect goods and food, yet light (decreasing transport costs), and can be easily formed into shapes, colored and designed for both loss protection and marketing purposes. Another driving factor is food distribution and protection. Plastic materials have brought us convenience through packaged items that are easy to grab and go and can also provide portioned food or drink when needed. In terms of food protection and packaging, there is not another material (yet) that has provided the advantages of plastic. However, some would say we have gone too far when foods that are naturally protected by their skins are wrapped in plastic. Plastic as a material for food packaging has had an economic advantage allowing companies to market effectively, designing appealing-looking and feeling packages, packages that prevent loss from store shelves, and making transportation less expensive with lighter weight materials.

But as the disposable culture for plastics grew, so did the concern over waste management. How would our waste management system handle a material it had never seen before? Even the industry knew this might be an issue as they discussed it at meetings where the information never saw the light of day in those time periods.[3] When plastics began to "leak out" of the system in the United States onto roadsides and into the environment, an extensive campaign to prevent litter began with an industry-funded organization called Keep America Beautiful (KAB). While it was important for the companies to provide resources to clean up a problem they were contributing to, it also misled the public into thinking it was their burden alone to keep the environment clean. Not only were companies not contributing resources upstream to infrastructure or prevention of waste and, therefore, litter in the first place, they also actively fought policies such as requirements for reusables and deposit-return schemes. They placed the responsibility for leakage and cleanup squarely on the shoulders of cities and community members. To this day, we continue to try to play catch-up with the problems that plastic pollution has caused.

This book covers plastics production and use, plastic waste generation and management, and environmental and societal impacts of plastic debris. We will cover policies and interventions to reduce the pollution caused by the use of plastic. In chapter 2, we will cover the rapid growth and use through time, including statistics on production. Since so many plastics are used quickly and discarded, waste management has also been an issue since its early use. In chapter 3, we will discuss challenges with managing plastic waste, from why recycling plastic has been so challenging to how a material that is so valuable to companies from an economic standpoint has so little value after it is used. Chapter 4 will discuss where and how much plastic has been found in our environment.

Next, plastics are not just a polymer but also contain additives and modifiers that can be released into the environment, as discussed in chapter 5. Chapter 6 will examine the environmental impacts of plastic pollution. Chapter 7 will address the impacts of plastics on different aspects of society. Because of the many headlines and echoed concerns about plastic pollution around the world, many governments acted. Chapter 8 discusses policies and regulations designed to address plastics. Chapter 9 will discuss interventions to reduce the quantity of plastics entering our environment and ocean, and the actions we can take to address this issue personally and collectively as we ponder what our future looks like with plastics.

In the 1999 movie *American Beauty*, a plastic bag danced around in a video with the characters discussing beauty. The scene wasn't really about how beautiful the plastic bag was, but it begged us to look closer at mundane things, everyday things, to appreciate the fact that there is an entire life behind objects. This book is about the entire life cycle of plastic things. We can't imagine our life without plastics. Yet, their contamination of our environment is just as evident—we see it nearly everywhere we go. Blame is often placed on the consumer or community member for pollution, but they are a part of a system with limited choices. This book can arm people with knowledge and information about plastics to make thoughtful choices when possible, influence change of the larger system, and spark inspiration.

2

PLASTIC PRODUCTION AND USE

It can be very difficult to imagine our lives without plastic as a material for use. Plastic can be molded into any shape, flexibility, and color and is used in sectors that have made tremendous progress, like transportation, medicine, and electronics. However, what society finds so useful about it, the sheer variety of things we can do with it, is precisely what often makes it so challenging to capture and manage after its use. When considering how to keep plastic out of the environment, we need to consider when, where, and how we use it. And the extent of this use became highly evident during the COVID-19 pandemic. During the early months of the pandemic, many of us were at home, often more than we had ever been in our recent lives, realizing the amount of plastic we use daily and how this use increased the waste we generated. People weren't necessarily generating more waste overall, but both the *types* of waste we were generating and *where* we were generating it were different. As of 2015, 8.3 billion metric tons of plastic had been produced globally since 1950, when only 2 million metric tons of plastic were produced.[1]

Plastics are a type of polymer, a large chain consisting of thousands of interlinked molecular units known as monomers. Plastics come from feedstocks like natural gas and oil, refined into components like propane and ethane. High heat applied to this feedstock will form the monomers propylene and ethylene, two common olefins (a double-bonded hydrocarbon) used to produce plastics and chemicals. Olefins are produced by fluid-catalytic cracking (oil) or by steam-cracking natural gas. The olefins ethylene and propylene are the building blocks of polymers and other chemicals (figure 2, polymers are highlighted in gray and other specialty chemicals are in white) and, after polymerization but before extrusion, will be in granular form. An extruder combines additives and the polymer, compounding and pelletizing the plastics for use to then manufacture their final plastic products.

The goal of cracking and producing olefins is not just for polymers, but numerous other chemicals as well. Typically, specialty chemicals generate higher profit for the petrochemical industry than refined gasoline, but there is only a finite amount of demand for these chemicals, like ethylene glycol. Any excess ethylene produced beyond that used for chemicals can be easily polymerized and

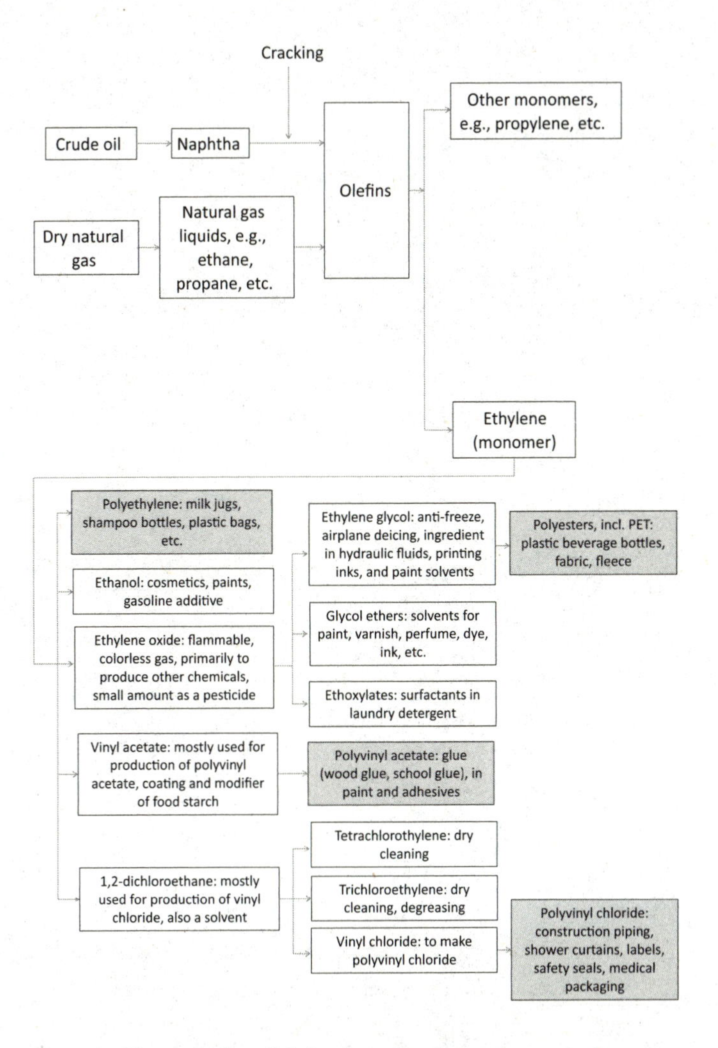

Figure 2 The production of olefins from oil and natural gas and subsequent related production of chemicals and polymers from the monomer ethylene.

Because of this, polymers' sheer existence and potential to add profit have pushed their use into more than what could be considered necessary.

made into plastic pellets for manufacturing, whether the plastics are immediately needed or not. Plastics are sometimes referred to as a by-product of oil or gas refining, since they are just one component of the petrochemical process and are made with excess monomers that are not used for higher-profit chemicals directly. Because of this, polymers' sheer existence and potential to add profit have pushed their use into more than what could be considered necessary.

While there are now over 5,300 polymer formulations commercially available, some of the most common polymers are polyethylene, polypropylene, polyvinyl chloride, polyethylene terephthalate, polyurethane, and polystyrene. Fossil-fuel-based plastics dominate the industry. Biobased plastics use alternate feedstocks such as wood pulp, sugars, crops, or fungi instead of oil or natural gas. Although biobased plastics make up less than 1% of the market, their demand is expected to increase.

Globally, 367 million metric tons of plastics were produced in 2020.[2] China produces most plastics (32%), followed by the North American Free Trade Agreement (NAFTA) countries of Canada, the USA, and Mexico (19%). The rest of Asia produces 17%, followed by the EU at 15% and other regions of the world at less than 10% each. Both polymer feedstocks (e.g., oil and gas) and polymerized plastics are traded globally. The US supplies 22.8% of the plastic manufacturing feedstocks to the EU, followed by imports from the UK, South Korea, Saudi Arabia, and China.[3] The most significant plastic processing imports to the EU are from China (33.8%), followed by the USA (11.7%). Plastics processing exports from the EU go to the UK (18.4%), followed by the US, Switzerland, China, and Russia.

The global production of plastics in 2020 decreased slightly from 2019, likely due to COVID. However, each year before that had been a steady increase since the recession in 2008, when production dropped fourteen million metric tons and bounced back the following year. The economy slowed down during COVID, but the use of plastic in several areas (e.g., medical and packaging) saw an increase in use, so the drop in production was minimal. The years 2020 through 2022 were "non-normal" during the height of the COVID-19 pandemic, with global supply shortages of many goods. Up to this point, in places like the US, consumption of items was at an all-time high.

Medical uses of plastic were critical for lifesaving uses in hospitals all over the world, but managing waste became an issue. Policies and regulations that had been already implemented or were about to be implemented to reduce plastic consumption (e.g., plastic bag bans, taxes, etc.) were officially rolled back. At that time, simple reusable mug, bag, and refillable product policies changed. Both governmental and industry policy changes resulted in more use of plastics because plastics are considered disposable. A great example is personal protective equipment and supplies like masks, gloves, and sanitation wipes made of single-use plastics. Demand for these plastic products outstripped supply at the beginning of the pandemic, but supply eventually caught up. Additionally, demand for packaged items and delivery of groceries, food, and other items increased, so more waste was generated at home. The increase in household waste resulted in a drop in commercial waste (waste at work) generation.

It might seem like durable items made with plastics take up the largest fraction of plastic produced overall, like cars, electronics, and building and construction materials. However, these are not the sectors of highest use for plastic globally; it is the packaging sector at 40.5% by mass.[4] That's right. Every product that is transported to its final destination for use is overwhelmed by plastic bagging, films, containers, tape, and padding (foam or air-filled plastic pillows). Building and construction is the subsequent highest use at 20.4%, followed by automotive use at 8.8%. Although it seems like every electronic device contains plastic, only 6.2% of plastic is used in the electrical and electronics sector. Household goods, leisure, and sports are a small percentage of use (4.3%), followed by agriculture (3.2%), and a combination of other uses at 16.6%. When considering where we use the most plastic, it is not surprising that it is in the packaging sector. Packaging provides for the transport and delivery of goods and products, but it enters the waste stream very quickly after goods are obtained or a product is consumed—and it often has no other use at that point. Think of all the packaging that accumulates in trash cans after holidays or birthdays. Plastic packaging often has no other value after its single use, which is why this is the sector of materials that most often end up leaking into the environment (chapter 4).

When considering how long a product stays in use, the US Environmental Protection Agency (EPA) sorts goods into three categories that use lifetimes in a material flow analysis to estimate waste quantities generated for a particular year. These categories are durable goods (stay in use more than three years), nondurable goods (stay in use fewer than three years), and containers and packaging (typically enter the waste stream the same year they are purchased). Examples of durable goods include appliances and furniture. Examples of nondurable goods include disposable diapers, trash bags, cups, utensils, medical devices, and household items such as shower curtains.[5] Packaging is what our food and other products come in at the store or when they arrive on our doorstep.

We now know that we can't start to reduce plastic pollution without a reduction of production. And plastic pollution starts at production, especially for surrounding communities (see chapter 7). In the past, plastics that were targeted for reduction were single-use plastics. Single-use was defined as something that had one use and could not be reused, like packaging, a disposable coffee cup, a beverage cup lid, a straw, etc. These were items most often found in the environment, in waterways, and in the ocean.[6] However, a new focus on plastic pollution reduction is to consider unnecessary, avoidable, and problematic plastics (UAPs).[7] There is a need for policy makers to be able to determine which plastics may be unnecessary, avoidable, or problematic in their context. For example, bottled water might be necessary in some contexts if clean drinking water is unavailable. Bottled water could be considered unnecessary in other places where refilling stations are free and available. Additionally, green plastic bottles, or any color instead of clear or blue, could be problematic for increasing recycling capabilities (e.g., clear and a few blue bottles are the only two colors that can be recycled bottle-to-bottle). As a community member, it is often difficult to avoid plastics—there is frequently no choice. And it can be a privilege to be able to avoid plastics when making purchases. Upstream and systemic change is needed, enabling more thoughtful decisions about where, when, and how plastics are used, if at all.

3

MANAGING PLASTIC WASTE

Trashed. Thrown away. Discarded. No longer "useful." When does plastic become "waste"? Some cultures do not recognize the concept of waste—the Tongan culture is one example. The word "waste"—and its related concepts, like landfill or trash bin—does not even exist in the Tongan language. What is waste? Is it something that is not (or no longer) useful? Or is it, as organizational guru Marie Kondo says, something that no longer "sparks joy"? And how should we get rid of something that we consider waste? In the context of plastics, these are challenging questions that we must grapple with as humans.

For thousands of years longer than we have been using trash cans and engineered landfills, we have been putting unwanted materials in our environment. Waterways such as rivers and the ocean historically were used as waste management systems because they transported waste away from our location. This was an act of management, when waste was initially all food waste or waste from our bodies. When metal and paper were discovered, they remained valuable materials because they could be recycled or easily burned, respectively. It wasn't until the age of plastics and "disposable" goods that solid waste became the *management* crisis it is today. Now, the traditional methods of managing waste have led to the plastic problem of today.

For this discussion, waste is defined as a substance or item that is discarded and either composted, recycled, disposed of, or placed uncontrolled in the environment.

Let's consider typical options for getting rid of waste generally, and items made of plastic, specifically:

- **Sell or donate the item** to someone else who might (still) find it of value.

- **Compost it** if it's biodegradable. Although plastics are carbon-based, they are often not biodegradable or compostable unless specifically designed to be so, as we'll explain later.

- **Recycle it**. This is what most people try to do with plastics. All the things that make plastics useful—they can be thin like a plastic bag, thick and rigid like a bottle, or any color of the rainbow—also make them difficult to recycle. Some plastics are recyclable, but some are not, so the majority end up in landfills, and, increasingly often, in unspoiled habitats and waterways.

The challenge of disposing of plastics is so overwhelming to consumers that some innovators are stepping in. For example, to address one of the most common questions asked of us, one of the authors of this book (Jambeck) and a startup at the University of Georgia developed "Can I Recycle This" (CIRT), a technology platform that can give on-demand, location-specific information about what can, and can't be, reused, recycled, or composted locally. Nonetheless, we are all now obligated to manage plastic properly, with the assumed responsibility placed upon us by companies when we obtain an item made or packaged with plastic.

Generation of Waste and Discarded Plastic

According to World Bank data, globally more than two billion metric tons of waste are generated annually.[1] Along with a steep increase in the production of synthetic materials, we have seen a resulting increase in plastic in the waste stream globally to an average of 12%. And from 0.4% in 1960 to 13.2% in 2017 in the United States alone. These percentages are all based upon mass, but to really visualize plastic in the waste stream, we can convert it to volume. A pile of trash often looks primarily plastic, but that is because plastic weighs very little compared to the space it takes up. For example, plastic bags, which are film plastics, spread out and look large because of their surface area but weigh very little. Rigid plastic containers are designed to hold products and, when empty, still take up the space reserved for the product.

Since the 1950s, more than six billion metric tons of plastic have been used and have become waste around the world.[2] Converting this quantity to volume, you'd be walking ankle-deep through plastic blanketing the entire state of Texas. With cumulative quantities projected to reach twenty-six billion metric tons of waste globally by 2050,[3] the need for reducing and managing plastic in the waste stream is only continuing to grow.

Management of Waste and Discarded Plastic

The US approached formalizing solid waste management by passing the Resource Conservation and Recovery Act (RCRA) in 1976, which required the US Environmental Protection Agency (EPA) to regulate solid and hazardous waste. "Open dumping" became prohibited and replaced by engineered and regulated landfills, composting, and recovery systems. Because of the RCRA and similar regulations globally, many solid waste management programs have been formalized. However, waste management remains a complex issue for many reasons, with environmental justice being one of the most pressing components.[4] No one wants a waste management facility in or near their community. As a result, facilities are often built in lower-income and historically marginalized communities.

A large, informal waste management sector operates around the world; about 20 million people count on picking, collecting, sorting, processing, and selling recycled materials for their livelihood. While the sector is significant in south and southeast Asia and Africa, waste collectors operate in greater Asia, North and South America, and parts of Europe as well. However, they are not often respected or recognized for their work. Waste pickers frequently have no protections for health and safety and no infrastructure for operation. Communities of waste collectors can live near dumpsites collecting valuable items under extremely hazardous conditions. This community has expertise in materials, recycling, and economics, and they are an essential stakeholder in policy and infrastructure discussions. Governments do not often account for their work. Still, they continue to provide waste collection and recycling, keeping plastic out of the environment despite all the challenges they face. In some cases, the informal waste sector has been able to form unions and negotiate with governments for recognition, pay or benefits, and improved working conditions and infrastructure. In Ho Chi Minh City, the (semiformal) independent waste collectors successfully worked with an NGO and negotiated a higher pay rate (paid by residents) for their daily collection and sorting activities, and an improved relationship with the government. There continue to be global efforts by many advocates and activists to improve working conditions and recognize this sector's contributions to keeping plastics out of the environment.

Reuse

If a plastic-based material retains some value, even if not for its intended (or initial) purpose, it can be reused, thus eliminating the energy cost of disposing or recycling it. For example, if you purchase a plastic yogurt or margarine tub, you can repurpose the empty and clean tub for storage. But you'll have too many tubs at some point, leading to clutter! Other formal reuse scenarios exist at stores and restaurants, and refill and reuse options are expanding thanks to organizations specifically focused on this work. Technology is facilitating this growth of reuse and refill with radio frequency identification (RFID) tracking, electronic payments, and more. These reuse and refill options prevent waste from being generated in the first place. The "best" waste is no waste at all.

Collection

Out of sight, out of mind. Most people want a system for quickly getting rid of their waste. Collection is often an expensive stage of the waste management process. The economics for collecting and hauling plastic waste can be tied to the waste generation of consumers who are not often disincentivized by the cost of managing plastics and other trash—the mindset may be, if you pay for a can to be collected each week, why not fill it up? In the US, a lack of curbside collection, more often in rural areas, can lead to increased illegal dumping and littering.[5] However, urban waste collection is also extremely challenging. For example, in New York City, bagged trash is just set out for trucks and workers to come by to collect in the wee hours of the morning to avoid disruption of daily life. Setting out bagged trash comes with challenges like animals (e.g., rats) getting into the garbage or liquids leaking out; however, lack of space for bins limits logistics. In Chile, small trash bags hang off of hooks on power poles or other posts so that they are not on the ground—and the collectors are familiar with taking them from their higher locations. And collection varies around the world, with many places collecting source-separated materials for recycling and recovery, as will be further discussed in the recycling section. But one of the most famous collection methods in the world is the singing collection trucks in Taiwan. The trucks arrive daily in neighborhoods announcing themselves with ice-cream-truck-esque music, like "Für Elise" by Beethoven. Community members bring trash and recycling to the trucks, where the truck operators meet them. They throw their own, already separated and bagged, waste into the trash truck. And then, the operators help them put their source-separated recycling into different bins on the recycling truck with precise separation that includes different bins for cooked and raw residual food waste. Collection in India may mean that a private waste collector, often called a "kabadiwalla," may come to your house to take certain materials you might even get paid for (figure 3). In a section of Beijing, China, there is a high-tech on-demand waste collection system, like Uber for waste.

Composting

The composting process often involves having microbial colonies process biodegradable carbon. Just like us, microbes feed on carbon-based material for energy; they need moisture (water), and if they are aerobic, they respire oxygen. Microbes in anaerobic environments process carbon into methane. Unfortunately, most plastics are nonbiodegradable because they are highly crystalline and their molecular chains are so long, making it difficult or impossible for microbes to digest them. Compostable plastics require formal composting facilities that can provide high temperatures (at least 40 degrees Celsius), control for oxygen, and curate microbial communities best suited for degradation. While some polymers that can compost or biodegrade at lower temperatures exist and are growing in popularity, their use is not yet widespread, and they still need to be managed properly. If they do not get composted and are landfilled instead, they will emit greenhouse gases like methane. When traditional plastic waste ends up in a compost pile and doesn't biodegrade, it contaminates the compost with small pieces of plastic.

Recycling

Globally, on average, we have only recycled about 9% of all plastic ever made.[6] How often have you thought or heard someone say, "I could recycle this, but I am not convinced that the items actually get recycled." What drives this general distrust of recycling? Cities bothering to recycle typically do what they can to get the materials recycled. There is potential money to be made compared to disposal, so outside of the trust, most are financially motivated. But recycling is not the answer to all our problems. One key component to recycling is local context. Since waste is collected and managed at the municipal level, local circumstances influence what can and cannot be recycled, and how recycling takes place.

Since waste is collected and managed at the municipal level, local circumstances influence what can and cannot be recycled, and how recycling takes place.

Recycle bins full of mixed items—bottles, cans, paperboard, paper, etc.—go to recovery facilities designed by engineers to separate what we all put in the bins. Some separation is done by hand. Other times the separation is automated via mechanical, optical, or magnetic methods. Japan and European countries (e.g., Norway) often have highly specialized and complex designed material recovery systems, although they often still require some source separation for collection. For example, in Oslo, there are different bins for paper, glass, residual, "dangerous waste," as well as return vending machines for polyethylene terephthalate (PET) bottles at stores. In Japan, there are reportedly up to twelve different categories for sorting waste in the home, although public waste sorting is simpler with cans, PET bottles being separated from combustible items. Although Japan reports an 84% recycle rate for plastics, 67% of that is combusted for energy recovery instead of being mechanically recycled.[7] Recent sorting technologies incorporate artificial intelligence and machine learning. The final step at a recovery facility is to compress similar plastics (milk jugs, for example) into "bales" of the smallest volume and highest density possible for efficient transport to a processing facility. The bales are cut open, washed, and then shredded at the processing plant. The shredded plastic is often used to make pellets, similar to the ones produced during virgin plastic production described in chapter 2. Recycled pellets still retain their original pigmentation; that's why mixed-color recycled plastics are usually dark gray. This process is typically known as mechanical recycling.

Recycling plastics has become a global trade issue as the World Trade Organization (WTO) encouraged trade in the 1990s, and China became the manufacturing hub of the world, but eventually, the country had enough. At the end of 2017, China implemented import restrictions so strict it became a ban on the import of plastic scrap for recycling. Before this restriction, China took over half the world's exported plastic scrap. And most exported materials were going from Organization for Economic Cooperation and Development (OECD) and higher-income countries to non-OECD and lower-income countries,[8] which were overwhelmed with the quantities coming in, especially the contamination that many countries included in their exports. This ban brought cascade impacts through the entire recycling value chain—material recovery facilities (MRFs) limited accepted materials or even closed for some time, and many people felt the impact on their kitchen recycle bins. Other countries followed China's lead, and, eventually, the Basel Convention was amended to require prior notice for the export of plastic scrap meant for recycling. And nearly 1 Mt of plastic scrap exported by the US was attributed to US leakage into the environment.[9]

In general, plastic is often "downcycled," meaning it is recycled into items that do not have the same use as the polymer's initial use, and this is often influenced by the previous additives and other chemical constraints that dictate what the recycled plastic is allowed to be made into. Beverage bottles can become bottles again, but often do not. Plastic bottles typically become new shoes or athleisure wear. Grocery bags and other film plastic sometimes get recycled into outdoor decking materials. Communities around the world have adopted creative ways of downcycling plastics. In Batangas, Philippines, where plastic regularly washes up on their shores, the community collects and separates the plastics, then melts them down and shapes them into stepping stones, garden edging, and other landscaping materials. In Kenya, recycled plastic is made into construction bricks.

Chemical and thermal recycling are currently expensive and energy-intensive processes that have not been able to be scaled. Thermal recycling uses heat, with or without oxygen, in processes of gasification or pyrolysis to transform the polymers into gasses and oils that could potentially be utilized for fuels. Chemical recycling breaks the bonds of the polymer to recover the monomers that can again be repolymerized. Although chemical recycling allows for a pure monomer to be utilized again, plastics, made up of polymers and additives, are a complex feedstock and need to be separated in this process. In addition, the energy consumption required makes these processes expensive. While development continues, chemical and thermal recycling technologies remain unproven.

It is easy to assume that more recycling, like in these examples, would reduce environmental plastic pollution. However, that's only true if there's value in recycling the material—as is the case for materials like aluminum or copper. For plastics, there's just too much of them, which reduces their value, and it's difficult to manage once they escape into the environment.

Landfills

Roughly 79% of all plastics have ended up in landfills or dumpsites around the world.[10] Landfills today are large, complex feats of engineering. In some locales, the peaks of the huge mounds eclipse buildings and natural hills. In its most basic form, the base of a landfill is a combination of geomembrane, a plastic liner about a half-inch thick, placed on top of clay. This composite layer is intended to capture any liquid that flows to the bottom of the landfill, called leachate. Similar to the sloped bottom of the basket holding coffee grounds that create a single stream of coffee that flows into a cup, the liner system of a landfill is sloped to one side, where the leachate is pumped out of the

a)

b)

Figure 3 a) The "singing" waste and recycling collection trucks in Taipei. b) Source separated collection of waste in Oslo. c) Women sorting imported plastic film by hand in Southeast Asia. d) Hand sorting recyclables at a kabadiwalla shop in India.

c)

d)

Figure 3 Continued

landfill. Leachate is not allowed to infiltrate the ground and is managed as wastewater. The landfill also generates gas, which is composed of about 50% methane and 50% carbon dioxide, which is actively collected through a network of gas pipes and blowers, and either burned at a flare to convert it to carbon dioxide or utilized as a fuel to create electricity or heat.

Engineered landfills are heavily regulated and must be sited after extensive environmental and geotechnical assessments. Plastics can be a challenging material in a landfill because they do not biodegrade and often end up in the leachate as microplastics.[11] Also, plastic films like grocery bags can be challenging because they are lightweight, get caught in the wind, and may blow out of the landfill. Landfills themselves can get fined for littering.

Combustion

Combustion is a process that uses heat to burn waste into ash and volatilized chemicals. Globally, 12% of the plastic that has been produced has been combusted and no longer exists in its original physical form on our planet.[12] Combustion of plastic is a very controversial waste management method, and understandably so. When waste is combusted and oxidized by the burning, it creates combustion byproducts. These chemical compounds are released based upon the temperature of combustion, the type of material burned, and other factors. Uncontrolled open burning of waste is illegal in many locations, but that is very hard to enforce on private land.

And in some cases, open burning is a chosen method of "management" in more remote areas around the world. Of the 2.4 billion metric tons of waste generated globally, an estimated 972 million metric tons (41%) of waste is openly burned at both residential and dumpsite locations. Open burning creates some of the most hazardous air pollutants, including the very small particulate matter of 10 and 2.5 microns in size, less than one-tenth of the width of human hair, benzene, PCBs, and dioxins in significant quantities. These emissions have not typically been included in air emissions inventories previously, and health impacts are substantial and not well understood.[13] This and other impacts of plastics will be discussed in chapters 6 and 7.

Industrial combustion is often regulated tightly because operations can impact emissions. Industrial combustion can create electricity and/or heat to be utilized, but if trash becomes a community energy source, an incentive for making waste can be realized. This is also known as "feeding the beast," which goes against many current movements to reduce waste generation rates. While combusting trash in a controlled and monitored setting reduces the volume of the waste by about 90%, there is still residual ash to manage. Ash captured in the gas management system is called fly ash, and ash left over after combustion is called bottom ash. Ash is a concentration of particles and elements that don't combust, like metals, and needs to be managed in a landfill system. While the ash landfill should theoretically not create any gases (all the usable carbon has been oxidized and should not biodegrade), the ash leachate can be quite high in metal concentrations. The leachate must be collected and managed in ways that protect human health and the environment.

Summary

The question of how we manage our plastic waste is complex, but the more we know about the plastic waste management process, the more we can work with all stakeholders to improve it. In some ways, managing our trash is not that different from our need to separate ourselves from our bodily waste. Environmental engineers have designed systems to make body waste "disappear" relatively quickly and easily. One fundamental difference between body waste and plastics is that most plastics are not biodegradable, and getting rid of them as quickly as possible is "wasteful" at best and harmful at worst. Engineered solutions to dispose of or recycle plastics create a false sense of security that we can endlessly use and discard plastic products. Historically, companies have avoided the cost of waste management related to the management of plastic waste related to their products—but more public awareness and government policies like extended producer responsibility (EPR) offer one way to change this.

4

DISCOVERY OF PLASTIC DEBRIS

The next time you are on an excursion to a national park, lake, or beach, take a moment to look around. You might encounter an errant bottle or a crumpled snack wrapper lying on the ground or in the water. You might consider it a nuisance or aesthetically displeasing on your getaway from the confines of modern technology and advancement. However, plastic trash in the environment is more than just an eyesore. It's a human-made artifact likely to outlive multiple generations of parkgoers. And believe it or not, plastic wasn't always in our waterways, soils, or falling from the sky. Yes, plastic rain is now real.

The 1950s brought in the era of large-scale plastic manufacturing along with minimal effort into handling the associated plastic waste products. This increase in production led to the rise in plastic not only in our landfills, but also on our roads, sidewalks, national parks, and waterways. Unlike an apple which can take days to decompose, some plastics can now maintain their form for hundreds of years. Unfortunately, it was unlikely that those manufacturing plastic or the general public utilizing these products in the early years were aware of the actual length of time these plastics could endure in these environments. This meant that plastics were treated like all other waste and could end up as litter in any environment. Therefore, it took multiple years before organizations and scientists took notice of this plastic buildup in the environment.

In 1956, an organization known as Keep America Beautiful began public relations campaigns to prevent littering. By 1971, plastic fragments were noted on the surface of the Sargasso Sea. And with time, more research slowly began to record plastic pollution on the ocean surface, coastline and seafloor. While roadside litter was a common concern to the American public, the outrage about plastic in the marine environment became much more prominent. This may be due to the idea that what you typically find on a walk down the road is usually something that is commonly used in your region. However, plastic waste in waterways can travel much further distances to new locations. And because the ocean is one large proverbial blender, even the most remote places like Midway Atoll, uninhabited islands in the middle of the sea, can now encounter international trash. So, the idea that local garbage is in contact with a global environment and people that never even generated that waste became problematic. Additionally, imagery surrounding the harm caused by plastic to marine animals like turtles, whales, and seabirds pulled on the public's heartstrings (chapter 6).

The Ocean Conservancy (formerly The Center for Marine Conservation) hosted the first International Coastal Cleanup in 1986.[1] The first cleanups typically gathered fragments of plastic and foam from coastlines; plastic cups, spoons, forks, and straws; metal beverage cans; foam cups; and glass beverage bottles. But it took some time for people to realize that this debris is not just primarily accumulating on shorelines. In 1997, Captain Charles Moore discovered the "great garbage patch" on the surface of the southern Pacific Ocean. Many of you readers likely have heard of the great garbage patch; however, the imagery you may have in your head is typical of an island of trash you can walk on and build a tiny home. However, the great garbage patch is not a walkable landmass. It is generally a location where the North Pacific Gyre tends to aggregate plastics and other floatable waste, including fishing gear, that has often started to weather and break into smaller pieces. Miles of floating debris and plastic can converge at ocean fronts, where two currents meet.[2] It took nearly two decades before scientists could begin predicting the sheer quantity of plastic used and disposed of in the environment (chapters 1 and 2).

Every new finding of plastic debris in our environment has led to an influx of organizations, nonprofits, and campaigns geared towards researching, cleaning, and remediating plastic waste in the ocean. In addition to the volunteer activities in cleaning up roadside or national park plastic debris, there are campaigns, nonprofit organizations, and social media influencers geared towards removing plastic waste from beaches and coastlines. As a result, the top debris items commonly found on the beach have evolved. By 2017, most of the top items found during the international coastal cleanup were plastic. These findings were not only plastics commonly used in households but also single use plastics. Top examples included cigarette butts (filters are made of plastic), food wrappers, plastic beverage containers, plastic bottle caps, plastic grocery bags, other plastic bags, straws and stirrers, plastic takeout containers, plastic lids, and foam takeout containers. Some of the most unusual plastics now found on beaches include dental floss picks, tampon applicators, band-aids, and pre-production plastic pellets. However, the most common human waste item on the coastlines now are plastics that can be smaller than a grain of sand. In 2005, this class of plastics was coined microplastics by noted plastic researcher, Dr. Richard Thompson.

Microplastics are pieces of plastic smaller than a pencil eraser (5 millimeters) that can be intentionally manufactured at those size ranges (primary microplastics) or be sourced from larger plastic debris that has fragmented (secondary microplastics). Nanoplastics are an even smaller size range of plastics discovered as they are typically smaller than 1 micrometer and go down to roughly 1 nanometer (a micrometer is roughly one-hundredth the width of a human hair, and one nanometer is one-thousandth the width of a micrometer). Primary microplastics include microbeads in body wash products, industrial paints, preproduction pellets, and glitter. The most common release of secondary microplastics comes from our own homes. Microplastic fibers (microfibers) are cylindrically threaded microplastics that can be found in our clothing, carpets, furniture, and single-use flushable wipes. Every moment we interact with these synthetic textiles or other plastic products causes some form of shedding or release of microfibers/microplastics (e.g., chewing straws, washing clothes or plastic food containers, vacuuming carpets, etc.). Even indoor and outdoor dust comprises a small fraction of microplastics and microfibers. It wasn't until 2010 that scientists measured microplastics in the outdoor environment quantitatively.[3] But by 2014, models had predicted over five trillion microplastics were floating on the ocean's surface.[4] Microplastics have now gained a bevy of public scrutiny as these tiny particles have been discovered in the highest points above sea level (Mt. Everest),[5] the deepest part on earth (Mariana trench),[6] and traveling to even the most remote islands in the world (Henderson Island).

a)

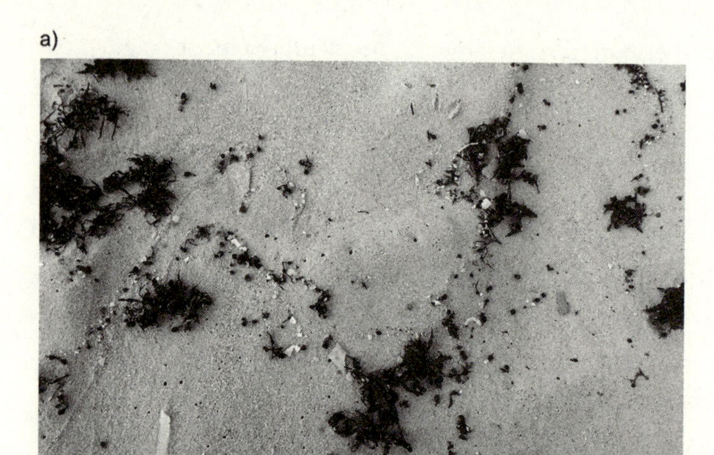

b)

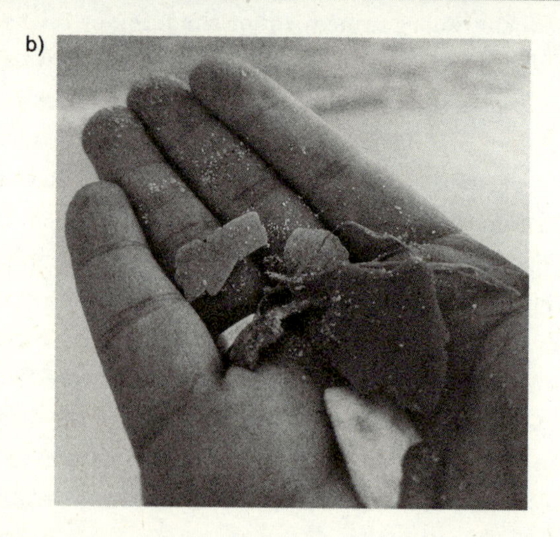

Figure 4 (a) Microplastic that has washed up on the shore in the strand line on the beach in Bermuda. (b) Close-up of plastic debris on the coastline.

There is currently no governmental action plan for mitigating microplastics flowing into our environment. The way we unintentionally deal with microplastics is an exercise of musical chairs. When our clothes are washed, or a synthetic wipe is flushed, millions of tiny microfibers flow into our water treatment systems. These water treatment plants usually can remove up to 97% of microfibers in solution. However, even a fraction as small as 3% of microfibers releases a relatively large quantity into receiving waters like our rivers and lakes.[7]

Additionally, these microfibers removed from our water treatment processes could be aggregated with other biological waste (sludge) and commonly applied as fertilizer for our agriculture. This creates a problem because these microfibers can then be released back into aquatic environments after rain events causing them to flow into stormwater runoff. And this means that the food we consume is growing in soils covered by plastic lining and mixed with an abundance of microplastics. Even plastics collected from the environment or water treatment facilities are disposed of in landfills. The plastic in the landfill can fragment microplastic into landfill leachate, where it is transported to a wastewater facility for treatment and continues the cycle of traveling microplastics.

Determining the fate of plastics in the environment can be tricky, but it all comes down to whether it will float, flow, or sink. And the basic principles determining this

Determining the fate of plastics in the environment can be tricky, but it all comes down to whether it will float, flow, or sink.

answer depend on the material properties of the plastic and its size. Light and small particles like microplastic and microfibers can easily float on the surface of the water or be carried by local and global air currents. For example, sea spray deposits 136,000 tons per year of microplastics on shorelines and into the air.[8] These microplastics that have been suspended in the air can travel regionally (fragments) or even globally (fibers). Western United States Parks have now predicted that over 1,000 tons per year of microplastics are deposited in remote wilderness areas from rain and dry deposition.[9] The aggregation of smaller plastic particles and microbial growth on the polymer surface will allow for the debris to become heavier. This change in weight means this plastic can sink and settle out of surface waters into deeper water and underlying sediment. Microplastic settling in the form of marine snow makes the ocean one large snow globe. Denser, larger plastic debris is more likely to sink and remain closer to the initial source of disposal. The determination of the density of a plastic relates to its polymer type: polystyrene (0.96–1.05 grams per cubic centimeter) plastics are similar to floating ice (0.9 grams per cubic centimeter) and are much lighter than sinking polyvinyl chloride (1.1–1.35 grams per cubic centimeter) plastics in water.

Plastic will interact with its surrounding environment no matter the location of this material in air, water, or soil. Plastic can be chemically degraded by water, air, and UV light exposure that will oxidize the polymer surface and create chain reactions that fragment the polymer chain. Physical degradation of these materials can be seen every day in plastic bags becoming brittle and breaking with the lightest touch, waves slamming into water bottles left on the beach, and tires on our cars losing tread. The most apparent visual of plastic deterioration can be seen in white and dark green outdoor chairs that, when left outside for too long, begin to lose their coloring, yellow, crack and eventually give out on us while we are sitting. The degradation of these polymers exacerbates the release of plastic fragments and chemicals associated with plastic into environmental systems.

5

CHEMICALS IN AND ASSOCIATED WITH PLASTIC

We all know that combining butter, eggs, flour, vanilla extract, baking soda, salt, and chocolate chips creates magical chocolate chip cookies. But did you know that a chocolate chip cookie is very similar to making plastic? Key ingredients added to a plastic make it look and taste (yes, some organisms can taste differences in plastics) different. Some common ingredients in plastic include monomers, catalysts, lubricants, flame retardants, biocides, UV inhibitors, and antioxidants.[1] The plastic building block, monomers, are like the flour, sugar, and egg that react and mold together with heating processes. Flame-resistant chemicals are similar to baking powder. Surfactants are the kinds of butter and oils added to keep the polymer from sticking while making it. Plasticizers are like the water added to a cookie to make the polymer more flexible and stretch out. Dyes and colorants are added to plastic to make a polymer aesthetically pleasing in the same way you would add chocolate chips or sprinkles to a cookie. Biocides are incorporated into plastic to keep microbial organisms from growing on the surface of the plastic (like how raisins are added to keep people from enjoying a perfect chocolate chip cookie).

The earliest formulations of plastics produced were not as stable as the products we interact with today. Some plastics, like Bakelite, would easily fall apart from exposure to hot water. Some would catch fire. Others were too rigid to do much with. Many simply weren't as pretty to look at. History museums preserving and displaying plastic artifacts from the 1900s use Raman spectroscopy and Fourier transform infrared spectroscopy technology to identify the polymer types and provide specialized care in maintaining the integrity of the artifacts. The most common methods for protecting aging plastic artifacts include keeping them cool, dry, in the dark, and sometimes away from oxygen. Some museums are now experimenting with applying chemical "sunscreens" to items that go on lighted displays.[2] However, plastics today have a variety of chemicals incorporated within the polymer and on the surface to aid in the material's functionality, aesthetics, and longevity. Now, plastic materials can be as flexible as a bendy straw, as durable as a water pipe, as resistant to wear and tear as a pair of yoga pants, and as aesthetically pleasing as glitter.

All chemicals are used for a particular purpose and can be slightly varied to make a different product. Examples include using a different chemical structure that is more environmentally friendly that can have the same function in the product as the previous version, adding other colorants to change the color scheme of the product, or adding new chemicals to protect the product from constant sunlight exposure. Therefore, even though a plastic is labeled polypropylene, it doesn't mean that it is the exact same product as another polypropylene, because different companies utilize various substances to process and protect that plastic during its use. Complete information on the quantity and type of chemicals used in an individual product are kept as proprietary business information. Privatized ingredient lists create a problematic continuous siloed feedback loop because each competing company has a different recipe to make the same product, leading to many types of chemicals associated with plastics.

In chemical commerce, there are roughly 350,000 substances registered globally. Many chemicals are used for agriculture, automotive, personal care products, and other uses. However, approximately 10,000 are used to make both synthetic and biobased plastics.[3] In comparison to similar materials, the number of entities associated with plastic takes up a sizable portion of chemical commerce: paper (5%) > plastics (3%) > wood (1%) > textiles (0.3%). Considering the total amount of material produced, 22% of the 820 million tons of global chemical demand per year is used for monomers of plastics, and 2.2% is used for plastic additives like plasticizers. The vast number and production volume of these substances highlight their importance in making the plastic products we see and use today.

The earliest glimpse into the chemical complexity of plastic can be seen in a plastic additive review paper published in 1975 by Deanin. Deanin defined plastic additives under the categories of "reinforcing fibers, fillers, and coupling agents; plasticizers; colorants; stabilizers (halogen stabilizers, antioxidants, ultraviolet absorbers, and biological preservatives); processing aids (lubricants, others, and flow controls); flame retardants; peroxides; and antistats."[4] Over the following decades, the categorization of these chemicals has evolved. Chemicals associated with plastics have been defined as 1) processing aids in the production and manufacturing of plastics, 2) plastic additives, and 3) manufacturing or environmental contaminants and degradants.

Processing aids include substances like lubricants. Lubricants are waxy or fatty compounds that can be applied to the machine parts and the exterior or interior of the polymer to aid in the molding process as it reaches temperatures above its glass transition state. These lubricating compounds have the added benefit of making a plastic item like a polyvinyl chloride pipe appear smooth on the surface and help equally distribute plastic additives like fillers and colorants across the polymer. Lubricants consist of 0.5 to 4% of the polymer weight in polyvinyl chloride. Instead of fat-derived lubricants, polyolefins like polyethylene use processing aids like fluoroelastomers that are incorporated to prevent the fracturing of the polymer while it is being melted by high heat. Typically, processing aids are compounds used during the melting and molding process of the plastic product. However, additional processing aids and polymer additives are added when recycling plastic materials into new products.

Plastic additives, also called functional additives, are used to improve plastic by taking on specific roles for the polymer. The prominent families are fillers, stabilizers, plasticizers, flame retardants, colorants, antistatic agents, and foaming agents. All these compound classes provide a function for the polymer: fillers strengthen the material; flame retardants offer fire resistance; stabilizers like antioxidants prevent aging and yellowing on the surface of the plastic; colorants dye the material, plasticizers increase flexibility; and antistatic agents reduce the potential for static electricity from contact with the item. While these are some of the significant classes of polymer additives, many other substances are used in lower production volumes that are equally important to get the final products we see today. For example, adhesives hold together the multilayer thin films used in products like your favorite bag of chips and sauce packets. While each compound in plastic usually serves one function, some chemicals can have multiple functionalities. For example, Bisphenol A, known as BPA, is a known antioxidant in plastics and serves as a monomer in making epoxy and polycarbonate plastics.[5] In general, the exact formula of which chemicals and in what quantities for making each plastic are not available to the public as they are proprietary. But we do know that additives can take up as much as 7% of the weight of the polymer.

Non-intentionally added substances, NIAS for short, are precisely as they sound. NIAS are chemicals not meant to be in the plastic; this includes impurities, degradation of other compounds used in making the plastic, residual monomers, and anything else that is along for the ride.[6] If we go back to our chocolate chip cookie example, NIAS is like finding clumps of flour, expired butter, or even hair in a cookie. These compounds are challenging because it then becomes difficult to backtrack the source of contamination or concentrations within the plastic when no one is reporting the use of an impurity in a product, let alone some companies don't even know they have this issue. Furthermore, once the plastic is recycled into a new item, the previous chemicals associated with the original object are now non-intentionally added substances to the new plastic product that are mixed with more processing aids and polymer additives.[7] As this is a class of substances that is widely under-studied and will increase in complexity, NIAS will be an essential part of the discussion for future plastic risk assessments.

Another group of chemicals unintentionally associated with plastic post-processing and manufacturing the plastic product are sorbed environmental contaminants. Commonly known as hydrophobic organic chemicals, these contaminants are chemicals that preferentially attach themselves to solid substances like plastics in the environment. This sorption process is just like how spaghetti sauce stains attach to plastic food containers or how the flavor of a cookie that has been left in the fridge too long starts to taste like the fridge itself. Examples of these kinds of chemicals include pharmaceuticals, pesticides, polyaromatic hydrocarbons, polychlorinated biphenyls, etc. While some of these sorbed compounds include toxic legacy chemicals long banned from production, the desorption of the chemical from the plastic into a liquid, solid, or gas environment becomes very difficult, along with overall degradation of the chemical.

However, chemicals associated with plastic during processing have every opportunity to leave the plastic the moment that it is made because they are not physically bound to the material. They can be emitted by vaporizing into the air, similar to the whiff of new car smell you get when opening a new package or getting new tires. They can also migrate into solids or liquids like our food and water. The time spans of chemical release can range from years to billions of years.[8] Release rates of the chemical depend on chemical properties, material properties, and the local environment surrounding the material containing the chemical. Chemicals closest to the surface have the easiest time to leave compared to those deeper in the polymer.

Release rates of the
chemical depend
on chemical properties,
material properties, and
the local environment
surrounding the
material containing
the chemical.

Chemicals that are more polar, less bulky in structure, and with small, low molecular weight tend to have a shorter time for release into the water. Plastics that are built like Swiss cheese tend to have more pores and free space within them, making them rubbery and amorphous. These large spaces allow for more accessible pathways for chemicals to escape. Accelerated release due to environmental factors can best be seen in surface layers of soils and water because of the extended UV light, temperature, biological organism abundance, etc. As more chemicals migrate out of the polymer from environmental degradation, the plastic becomes more brittle and worn. This cyclical process creates new cracks and extends the pathway for more chemicals to be released from the polymer.

As chemicals migrate out of the polymer, they can also undergo transformations based on environmental conditions. Chemical transformation can occur by exposure to heat, light, air, water, or biological organisms. These processes usually cause the addition of oxygen or water and can fragment the chemical or polymer chain leading to the loss of monomers. Some of these chemicals, like perfluorinated compounds, can persist for long periods in the environment and accumulate in soil or body fat deposits of organisms. The release, occurrence, and transformation of these chemicals call into question the potential of these compounds to be toxic substances before, during, and after the use of plastic.

6

ENVIRONMENTAL IMPACTS OF PLASTIC

Plastics are considered a novel entity on our planet, as they are anthropogenic and have only existed on a short geological time scale. Plastic debris in its macro, micro, and nano size scales has been emitted worldwide and has traveled nearly the entire planet. This movement of plastic waste has allowed for accumulating chemical and biological constituents on a polymer surface while simultaneously releasing polymer-associated contaminants into the surrounding environment. Due to the magnitude and scale of global plastic pollution, plastic recovery is no longer readily reversible, as it would be impossible to retrieve all plastic or associated chemicals produced since the 1900s. Because of this ubiquity in the environment and the still unknown large-scale disruption to organisms and geophysical processes, plastic pollution is considered a potential planetary boundary threat.[1] To explore the impact plastic pollution has on the planet, it is vital to understand the influence plastic has on wildlife populations and biodiversity, environmental air and water quality, and biogeochemical processes (figure 5).

Wildlife Interactions

Media and even plastic pollution campaigns often use imagery to convey the physical impact of plastic pollution on wildlife. Iconic videos include Samaritans dislodging a disposable plastic straw from the nose of a turtle, freeing birds from fishing nets, taking off plastic bags wrapped around a sea lion's neck, and cutting off plastic six-pack can holders wrapped around a dolphin's mouth. Unfortunately,

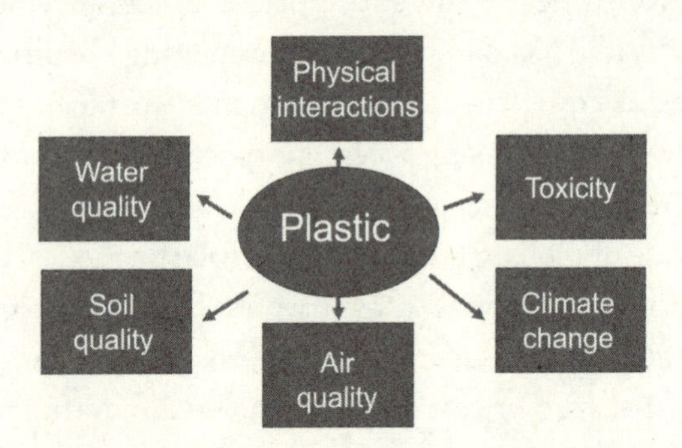

Figure 5 Diagram of the many ways that plastics can interact with the environment.

due to the large quantity of plastic in the environment, wildlife is increasingly observed becoming entangled by nets, ropes, fishing lines, and other plastic debris. Ghost gear, such as nets, traps, and strings, are derelict fishing gear that captures fish and other marine life. Common injuries from plastic debris entanglement include limited mobility, physical scarring, strangulation, loss of limb, and sometimes death. Loss in marine biodiversity is harmful to the population dynamics and the food web and can lead to reduced fish populations vital for fishing industries and communities that rely predominantly on fish for sustenance.

Marine life has been observed using plastic debris to form their habitats. Birds have stuffed their nests with plastic bags and netting to insulate new eggs before they hatch. New habitats for marine species have been observed underneath marine plastic docks that host species of barnacles, algae, microbes, and fish communities. But while plastic debris can create new habitats, it can also destroy habitats for others. Phytoplankton and coral reefs have experienced habitat loss and displacement when plastic debris and particles cover the ocean's surface and prevent photosynthesis, a necessary function for their life to continue. Marine or microbial communities can also attach themselves to plastic surfaces buoyant enough to travel with the currents to new locations worldwide. The large-scale movement of marine debris not only displaces the organisms but also introduces these species as invasive to the new environments they inhabit.

The earliest interaction of an organism with plastic was recorded in 1969 when seabirds were observed ingesting plastic.[2] Decades and many research articles later, plastic consumption has been reported within 1,288 marine species and 277 terrestrial and freshwater species.[3] Even today, households provide pet birds, dogs, and cats with bright, colorful plastic toys to chew and sometimes accidentally consume. Ingestion of plastic by wildlife is proposed to occur if plastic is in the environment, the animal is hungry or lacking some form of nutrition, or their prey resembles plastic. An example of this is loggerhead sea turtles, often mistaking plastic bags for jellyfish. Birds target plastics due to the variety of colors they display, such as their attraction to bottle caps.

Additionally, chemicals released from plastics can impact sensory signals and lead to the increased appeal of an organism to plastic. Hermit crabs and sea anemones have been observed to consume plastic like chewing gum until the "flavor" wears out. Microbial communities utilize chemical nutrients on plastic surfaces as an energy source and prefer aged plastic to form large gatherings on the polymer surface. Microbial colonies can also engulf the surface of smaller plastic particles and lead to the formation of biocoronas. This biofouling alters the biological interactions of the plastic with other organisms like sea urchins. This relationship with plastic eventually leads to the mineralization and eventual biodegradation of plastic into fragments and microplastics. But to continue this cycle of consumption, microbial growth can, in turn, signal for larger organisms to consume the plastic.

Implications of ingestion of plastic include nutritional dilution in the gut as organisms no longer consume as much of the typical nutrients as they would without plastic in their diets. Dietary dilution is defined as animals with plastic in their stomachs consuming smaller quantities of food or feeding frequency compared to animals who have not consumed plastic. This result could be a decreased gut volume available for food and plastics affecting hunger and satiety. In addition, depending on the size of plastic debris, ingestion can cause choking or clogging in the gastrointestinal tract. Examples of increased mortality rates related to plastic consumption have been observed in elephants grazing on plastic in landfills[4] and camels accumulating plastic in their "polybezoar."[5] And if you are wondering what a polybezoar is, it's sort of like all the swallowed gum you ate as a child that a parent told you would never leave your digestive tract clumped together. (P.S. Most of that gum is also plastic.)

Particle size determines the uptake, body retention, and effect plastic has on the organism. Plastic fragments, while incredibly small, can be ingested and inhaled from the air. And if the particle is in the nanometer size range like a nanoplastic, it would be small enough to migrate into the bloodstream and accumulate into different tissues. The harm from its size differs by its location in the body. Nanoplastics in the intestine can increase inflammation and oxidative stress. Microplastics in the liver can increase oxidative stress and lipid metabolism disorders. However, both microplastics and nanoplastics have the potential to induce neurotoxicity, downregulated gene expression, and decreased reproductive health.[6]

Chemical Toxicity

In chapter 5, we uncovered that chemicals and materials are added to a polymer to enhance the properties of the desired product and that plastics can release these substances into various environmental and biological matrices. Chemicals released from weathered plastic can potentially be ecotoxicological hazards, including carcinogenic, persistent, mutagenic, bioaccumulative, neurotoxic, endocrine disrupting, or aquatic toxicants.[7]

A typical case study of harm caused by polymer-associated chemicals to wildlife is via analysis of the chemical Bisphenol A, also called BPA. BPA is a monomer and antioxidant used in plastics that are shown to mimic or block a body's hormones. BPA has been proven to be able to be released from plastic into various environmental matrices and is exacerbated by the degradation of plastic products like food and beverage containers. BPA has been detected in wastewater, marine and freshwaters, agricultural biosolids, landfill leachates, soils, and the atmosphere. In addition, animal studies have found that high concentrations of BPA impacted sexual and neural development in mammals, birds, amphibians, aquatic reptiles, and fish. Some effects included depletion in attraction to the opposite sex mate, sex reversal of males to females, and declined sperm quality.[8] Because of these many findings and public concern, many products have removed BPA from the plastic used in their products and include the label "BPA free."

Depending on the chemical associated with the plastic, the gut is a much more energetically favorable environment for the release of the chemicals from plastic compared to the water in which the plastic might have been floating in before ingestion. For example, the presence of nonylphenol and triclosan associated with ingested microplastics was found within the tissues of lugworm guts after ingestion. Those lugworms subsequently had a reduced ability to remove pathogenic bacteria, diminished ability to engineer sediments, and a mortality increase.[9] Ingestion of plastics has also been found to increase the liver concentrations of metals like aluminum (Al), manganese (Mn), iron (Fe), cobalt (Co), copper (Cu), and zinc (Zn) in the Slender-billed prion seabird.[10]

Databases have found that 68 of the 906 chemicals likely used in plastic packaging are listed as environmental hazards, and 34 of the 906 chemicals as known or potential endocrine-disrupting chemicals like BPA.[11] However, most polymer-associated chemicals have not been thoroughly tested for their potential toxicity. There is also concern about the toxicity of chemical mixtures released from polymers due to additive, synergistic, or antagonistic effects. Laboratory studies have found that some plastic extracts from consumer products can induce a toxic response as a chemical mixture. Polyurethane and polyvinylchloride extracts are the most toxic plastic-type, containing hundreds to thousands of chemicals. Even extracts of bioplastics have been shown to contain complex mixtures of chemicals that can cause oxidative stress, baseline toxicity, antiandrogenicity, and estrogenicity.[12] In the environment, it is more difficult to determine the environmental impact on species from chemical mixtures already present and those mixtures being released from plastics.

Similar to the chemicals intentionally placed into plastics, environmental transformation products of these chemicals have the potential to be harmful. For example, a chemical known as 6PPD (*N*-(1,3-dimethylbutyl)-*N'*-phenyl-*p*-phenylenediamine) is used in tire production as an antioxidant used to protect the rubber from cracking and degrading when exposed to air. However, particles that come off from losing tread on tires accumulate on roadsides and are washed into our freshwater streams and lakes during storms. The presence of tire rubble in the environment results in the release of 6PPD and environment transformation via ozonation into 6PPD-quinone. This compound, 6PPD-quinone, has been directly linked to acute toxicity and death of Coho salmon in the Pacific Northwestern United States after exposure to urban runoff.[13]

In addition to water quality, scientists are exploring the impact of plastic on the soil environment. The majority of plastic produced stays on land during use and after use in landfills and the terrestrial environment. In the UK, up to 200,000 credit cards worth of microplastics are dumped monthly onto agricultural land coming from wastewater treatment facilities.[14] Questions still being pondered include the uptake of micro and nanoplastics into plant life and accumulation in the vegetables and fruits we consume, changes in the nitrogen and carbon cycling in soils, and changes in the microbial communities and microfauna in soils. Even the very presence of microplastics in sand creates temperature variations on the ground that loggerhead turtles lay eggs in and has been shown to alter the sex of the species.[15]

This warming of the surface due to increased plastic concentrations can also be directly related to the impact of plastic on climate change. The production of plastic requires raw materials harvested from the earth and leads to methane leakage into the atmosphere. The process of distilling and manufacturing the final product will release greenhouse gases. During use, plastic is known to degas greenhouse gases methane and ethylene, especially when exposed to sunlight. Plastic disposed of in the environment and allowed to fragment into microplastics can cover the ocean's surface and affects phytoplankton's ability to complete photosynthesis, which can alter the ocean's ability to store carbon.[16]

Environmental Quality

Plastics and their associated chemicals introduce an array of contaminants to environmental matrices like water, soil, and air, and call into question their safety. Ecological quality of air, water, or soil is defined as a chemical (PH, dissolved oxygen, etc.), physical, and biological (algae, phytoplankton, etc.) measurement of suitability for a particular use. Suitability includes clean water for drinking, air for breathing, or soil for plant growth. Safe conditions include clean air for birds to fly in, water for organisms to live in, and air and water for agriculture. An example of regulation for water quality in the United States can be seen in the elimination of DDT, which was thinning eggshells in birds, and the limitation on mercury use after heavy mortality of fish and birds in the water. In addition, the Environmental Protection Agency reframed air quality standards after intense acid rains in the northeastern United States. Evidence has shown the physical and toxicological impacts of plastic not only have the potential to impact an individual but entire food webs. However, there is currently no benchmark for the amount of plastic and microplastics or polymer-associated chemicals allowed in air or water, even though some government agencies are fervently working to address this.

Furthermore, research is still early in quantifying the scale of pollution of these contaminants in the environment and determining the systemic consequences of these quantities of plastic. Therefore, it has been difficult for policymakers to decide how to determine risk and limitations adequate to limit plastic pollution. And while large-scale impacts have not been entirely resolved, there is enough urgency for research and policy changes to be implemented to protect the environment.

7

PLASTICS' IMPACT ON SOCIETY

The harm caused by plastics to wildlife, as highlighted in chapter 6, is informative to how humans may be impacted by the plastic that fills their daily lives. Many individuals and environmental justice groups consider humans a part of Nature and that all people, human and non-human, are intertwined. However, this chapter dives deeper into how human society benefits but is also hindered by plastics.

Human Health

Plastics were initially perceived as an "inert" material and therefore considered minimal concern to human health. The discovery of microplastics and plastic additives in food products began to alter that story. Microplastics come in various sizes that have the potential to be transported globally around the world and internally in the body. These fragments can have sharp edges or be charged on the surface, which can directly impact neighboring cells in the body. But in an exciting twist, the medical community has been investigating for years the use of micro- and nanopolymeric spheres as a pharmaceutical method for drug delivery which might even inform us how similar micro-sized polymers from sources like food packaging will interact with the body on a cellular level.[1] Many scientists and organizations have called for an investigation of the direct impact of plastics on the human body. However, research is years behind the media frenzy for answers. Why? Because it would be very difficult (and ethically questionable) to do a controlled dosing study of plastics on humans. Scientists have to rely on observations, correlations, animal studies, and mathematical models to create human health risk assessments for plastics.

Some of the first human health-related research observations discovered microplastics in the drinking water, food, and air we breathe. We can witness this in the media headlines for microplastics found in beer, salt, seafood, tea, honey, soft drinks, bottled water, and tap water. The evidence of plastic particles in what we eat and drink indicates that microplastics are a regular part of our diets. Models have now estimated that adults are consuming and inhaling more than 70,000 particles a year.[2] Further observations confirmed this by detecting microplastics in adult and toddler feces. But some key age difference is that there are ten times more microplastics excreted from young children compared to adults. Outside of life-saving plastics used in the medical field, childhood exposure to microplastics raises scientists' concerns for future generations. One proposed explanation for increased microplastic exposure might be that young children interact with a larger quantity of plastic via baby bottles, feeding utensils and containers, and toys. For example, polypropylene baby bottles have been shown to release thousands to millions of microplastics into the heated formula. These findings introduce more significant concern for human health as fecal microplastic concentrations are correlated with irritable bowel disease (IBS).[3]

Depending on the size of the plastic fragment consumed, some microplastics can migrate into the bloodstream and accumulate in different tissues. For example, studies of pregnant rats that inhale microplastics have found plastic particles rapidly migrating from the lungs into the heart, brain, and portions of the fetus.[4] Further confirmation of this phenomenon is observed in studies that have found microplastics in human blood, lung tissues, digestive tract, and the placenta.[5] This is concerning because even smaller size ranges of plastics, nanoplastics, are also predicted to be able to enter human tissues and produce inflammation. A direct example of these risks can be seen in the sleep apnea machines recalled in 2021 for the risk of causing asthma, lung irritation, or cancer.[6] This recall was due to the polyester-polyurethane foam that could degrade with

Models have now
estimated that adults
are consuming and
inhaling more than
70,000 particles a year

use over time. The degraded foam could then release chemicals and microplastics directly into the air the users were breathing at night. Previously in 1975, occupational exposure of workers to fibrous and nonfibrous synthetic particles in factories was linked to outbreaks of interstitial lung disease and suggested to increase the risk of lung cancer.[7] While correlations from occupational exposure epidemiology studies do not reflect exposure to the general population, this information can be helpful in identifying hazards after elevated long term exposure to microplastics.

Chemical Exposure

Additionally, there are human health concerns from chemical exposure originating from plastic particles acting as a vector for released adsorbed chemicals and chemicals added to the polymer during processing. Some of the most common plastic-associated chemicals include Bisphenol A (previously discussed in chapter 6), styrene, perfluorinated compounds, and brominated flame retardants. Styrene is a monomer of polystyrene foams associated with the potential to cause cancer and is found to migrate into the food it contains.[8] There is an increased risk of styrene migration with increased fat content and temperature. Additionally, it is very difficult to successfully recycle polystyrene containers and products. Because of these findings, legislation has increased to eliminate foam food takeout containers and polystyrene egg cartons, including denoting polystyrene as "problematic" in several frameworks, one being the US Plastics Pact.[9] Perfluorinated chemicals like PFAS have been titled "forever chemicals" due to their longevity in the environment and the body. They have been linked to cancer and other toxic responses but are still used as surfactants and to form non-stick surfaces for food wrappers at major fast-food restaurants, carpets, furniture, nonstick pans, and other commercial products.[10] Brominated flame retardants were a class of chemicals used to inhibit the flammability of a product. Brominated flame retardants were commonly applied to couch cushions, synthetic textiles, and electronics. These persistent chemicals have been discovered in remote parts of the world, such as the tissues of arctic animals, closer to home in our house cats, and can bioaccumulate in breast milk. Brominated flame retardants are of concern due to the potential for these chemicals to cause cancer and are linked with cognitive decline and reduced fertility.[11]

Based on these observations, studies on model species and mathematical models predict that harmful chemicals have the potential to be released from microplastics into the stomach and intestine. Furthermore, these chemical mixtures are correlated to mechanisms that influence fat buildup in cells.[12] In addition to the possible hazards of the physical plastic particles and the chemicals associated with these polymers, plastics can also harbor pathogenic bacteria. However, this has not been as widely investigated. Conversely, to this point, plastic can also protect food from both oxygen and bacteria, which will prevent bacterial infections in food products like Salmonella and other foodborne diseases.

Health impacts from plastic use and fragmentation are challenging to determine cause and effect because there are other pollutants we are exposed to daily. While correlations of plastic pollution or plastic-associated chemical exposure to harmful effects exist, it is difficult to directly determine the level of impact and relationship to diseases and ailments. However, without more research and risk assessment studies, there is no definitive answer to every possible consequence of plastic ingestion and inhalation into the human body. The World Health Organization concluded in 2022 that research evidence for impacts of microplastics and nanoplastics on human health is insufficient to determine harm. However, they also stated that this lack of evidence does not mean that plastic particles are safe.

Environmental Justice

In addition to the concern about the health impacts of plastic exposure on the population, there is a significant environmental justice issue surrounding plastic production and pollution. If we think about environmental justice from the perspective of distributive justice, it would be defined as the inequitable distribution of harm to communities, usually highlighted by differences in geography, race, ethnicity, education, or socioeconomic factors. An example of this type of inequitably distributed harm from plastic is shown in fence-line communities near where oil is extracted, refined, distilled, and manufactured into plastic products. These communities are disproportionately exposed to large quantities of toxic chemicals released into the neighboring air and water. Cancer Alley in Louisiana, USA, in particular, is one place where communities are exposed to these pollutants and are burdened with higher incidences of asthma, cancer, and other health disparities.[13] Additionally, the lack of affordable and accessible alternatives to the abundance of plastic packaging often leaves communities with no agency to make their own choice in plastic exposure. This can be noted in food deserts historically prevalent in neighborhoods of color only containing gas stations, fast food, and dollar stores as substitutes for grocery stores. Because of this, most food is plastic wrapped, and more products have higher ratios of plastic packaging to product. However, plastic-wrapped food also means that food takes longer before it spoils and allows for the transport of food much longer distances around the world.

At the stage of plastic disposal, fence-line communities can also be surrounded by incinerators and landfills that continue to expose their families to more plastic and chemical pollution. Plastic imported to countries like Vietnam, Turkey, and India, is placed in a catch-22. This plastic waste provides work for waste pickers that filter through the trash for high-value plastics like polyethylene terephthalate (PET). However, these workers and communities in greatest proximity to plastic waste are at risk of potentially getting sick, and the "low value" trash tends to pollute their communities and waterways. To counteract the accumulation of waste on limited available land and in waterways, some communities openly burn their trash as a part of their daily life habits. Open-burning plastic is linked to asthma and cancer as it commonly releases microplastics in sizes readily available to be inhaled into the lungs and releases potentially harmful plastic additives and their degradants like dioxins. Incineration also causes the release of greenhouse gasses into the environment, which exacerbates climate change. If we assume that plastic production continues to increase, by 2050, plastics could contribute around 13% of the global carbon budget.[14] The plastic lifecycle's contribution to climate change will thus exacerbate the number of climate refugees forced to leave their homes due to rising sea levels and temperatures.

Some locations are not intentionally importing trash but are seeing the massive accumulation of plastic debris and particles in their waterways and lands, such as Midway Atoll, Canary Islands, Henderson Island, Easter

Island, and Tonga. The accumulation of plastic waste on Indigenous lands infringes on Indigenous sovereignty, defined as "Indigenous peoples' rights to govern their lands, waters, and lives."[15] Water and ice in Arctic locations, including Inuit Nunangat (Inuit homelands), contain concentrations of plastic particles likely sourced from both regional and long-range transport.[16] This also includes microplastics found in fish used for sustenance in local diets. Approximately three billion people in the world rely on fish (farmed or wild caught) as a primary protein source that can now contain plastic particles.

Internationally, governments are working on setting limits to allowable quantities of plastic/microplastics in the environment and drinking water. Still, some argue that any waste in the environment is not safe. Developmental justice is defined as the right of an industry to continue its growth trajectory and not be impeded by its economic goals. So if we apply a developmental justice model framework, enterprise and capitalism will continue to make plastic for progress, indicating that harm to previously highlighted communities will continue to be perpetuated. New policies implemented locally, nationally, and globally are vitally necessary to create real changes that address the harm caused by plastic pollution.

8

PLASTIC POLICIES

Collecting plastic and paper bags since 1975, Heinz Schmidt-Bachem of Germany holds the record for keeping 150,000 of them,[1] but if you opened a cabinet, drawer, or pantry in your kitchen, how many plastic bags might you find? We use, and try to keep and reuse, so many plastic bags that companies make holders that stick to cabinet doors, organize drawers, and other containers that bags can be squished into, specifically to contain the multitude of plastic bags that seem never-ending in our lives. While we make every attempt to reuse them, the uses do not typically keep up with the sheer quantity we can be given. Plastic bags use is estimated to range from billions to trillions per year, equaling about one to two million per minute, the most ubiquitous consumer item in the world.[2]

Some people still recall a time without plastic bags. They were not invented until 1965 when the Swedish company Celloplast patented the first plastic bag to save trees from being made into paper bags. Use quickly spread to the USA and other countries when in 1979, plastic bags held 80% of the bag market in Europe. In 1982, two of the biggest supermarket chains in the USA, Safeway, and Kroger, switched to plastic bags. By the end of the 1980s, plastic bags had become the norm throughout the world.[3]

Besides the sheer overwhelming quantities, it didn't take long for people to see the downside of so many plastic bags. The first documented entangled plastic item after fishing gear was reported to be a plastic bag captured by a Continuous Plankton Recorder off northwest Ireland in 1965,[4] and the deepest litter ever found in the ocean was a plastic bag at 10,898 meters depth in the Mariana Trench. Plastic bags are challenging to recycle as they cannot be put in the curbside bin, get tangled in the conveyor and screening systems at material recovery facilities, blow out of trash cans, trucks, and away from landfills easily, and clog drainage systems causing flooding.

It's no wonder the earliest and most frequently enacted plastic policies pertain to plastic bags. The first plastic bag ban in the world came from Bangladesh in 2002, recognizing the harm they were causing to drainage systems in the country. The first tax on plastic bags was implemented in Denmark in 1994. It is estimated that Danes use about four single-use plastic bags a year while people in the USA use three to five hundred. Other countries have followed these examples. As of July 2018, the United Nations Environment Programme (UNEP) documented that out of 192 countries reviewed, 127 have enacted some form of policy to address plastic bags. In other countries, like the USA, where no federal policy was in place and with much plastic use and waste management activities occurring at the city and community-level scale, regulations often first occurred at that level before becoming state or nationwide. The Surfrider Foundation mapped over 1,000 existing city or state plastic reduction policies within the USA, including bans or taxes on plastic bags, straws, and expanded polystyrene.[5] But it has been a policy war, often pitting communities against industry through plastic bans versus preemptive bans and small business against big business. Six plastic bag manufacturers sued reusable bag company, Chico Bag, in 2011 for data misrepresentation that was actually accurate and the bag manufacturers lost and had to make labeling changes.

The number of country-level plastic pollution policies has been increasing over time. The Plastics Policy Inventory held at the Nicholas Institute for Environmental Policy Solutions at Duke University houses a large subset of policies adopted from the early 2000s until 2020 to reduce plastic pollution.[6] Patterns in these policies emerged through their analysis. A large subset of the policies targets more than one plastic type using bans, with macroplastic items being the most common and 89% of the bans and regulations targeting plastic bags. The most common policies include either a ban, fee, or return policy on plastic items. A ban is described as a limitation or restriction of an item sold or given out at stores. A ban was most commonly used for items like foam packaging and plastic bags, known for their inability to be recycled, their longevity in the environment, and their potential deleterious impacts on human health and the environment. A tax or fee is used to encourage the reduction of the amount of plastic used by consumers but still make it available, and it is most often used in the context of bags, where fees are in the range of five to ten US cents per bag. After the introduction of a plastic bag tax or fee, consumers may have to purchase bags for the applications for which they were using the free bags. However, there has been some evaluation of policies two years or more out where a reduction of more than 50% of plastic bag use was observed.[7] With the plastics industry lobbying efforts, some states, eighteen as of 2022, have put into effect preemptive bans of policies to ban or tax plastic items. So in those locations, no bans or taxes on items like bags can be enacted.

Education and outreach were found to enhance compliance with either bans or economic incentives, and funds from taxes, for example, can be used for these campaigns. Interestingly, education and outreach alone were unlikely to change behavior.[8] Bottle deposit-return schemes are also commonly implemented and often for more than just plastic materials. Historically, deposit-return schemes were used for refillable containers like glass milk and soda bottles. Bottle deposit systems typically give the consumer money back for the bottle ranging from five to ten US cents. Compared to what leaks into the environment, policies like a deposit return scheme on plastic beverage bottles result in 40% fewer bottles ending up in the environment where these policies are implemented.[9]

The only national ban in the USA is a ban on plastic microbeads in cosmetics, which had a direct input of microplastic to waterways. This ban was passed with bipartisan support in 2015 and implemented in phases from 2017—2019.[10] Similar prohibitions exist in Canada, France, India, New Zealand, Sweden, Taiwan, and the United Kingdom. Another regulation implemented in the USA regarding plastic, as of March 1, 1994, is that all plastic six-pack rings must be photodegradable, known as the Degradable Plastic Ring Carrier Rule. While it did mean that seals and other animals being killed by plastic six-pack rings would be saved, it also meant that any six-pack rings ending up in our environment would degrade into plastic fragments faster than ever before. In early 2022, California became the first state to enact a Statewide Microplastics Strategy, put into place by the California Ocean Protection Council.[11] This strategy has a two-pronged approach, one of them being proactively pursuing solutions the state can act upon immediately while the scientific knowledge of microplastics specific to California further develops. And then secondly, to invest in and advance scientific understanding of microplastics to develop and refine future solutions.[12]

But despite finding plastics just about everywhere we look, concentrations of plastics in our environment are not regulated at any level. Plastic trash items would be included in US Total Maximum Daily Loading (TMDL) regulations, but these are for all trash and not specific to plastic in our environment. TMDLs are also not necessarily implemented as a strict limit, they can be voluntary guidelines and are often aspirational goals for zero trash in the specified waterway. While there have been previous calls for plastics to be considered a hazardous material to qualify for cleanup as a pollutant,[13] these proposals have not been advanced. However, in 2021 Canada classified plastics as a "toxic substance" under the Canadian Environmental Protection Act of 1999. This ban means Environment and Climate Change Canada (ECCC) has the authority to enact regulations to ban or limit the use of certain plastic products and implement pollution prevention plans and environmental codes of practice. Items previously identified by the government for bans are checkout bags, cutlery, food service ware made from or containing problematic plastics (as outlined in chapter 2), ring carriers, stir sticks, and straws. Many countries around the world have been more proactive and aggressive than the US in addressing plastic pollution with policy. In Kenya, the highest penalty for possessing a plastic bag exists: imprisonment of up to four years plus a fine of $40,000, and countries on the African continent are leading the world in plastic bag regulations, with thirty-one bans passed in sub-Saharan Africa alone.[14] And Indonesia was the first country to write a National Plan of Action for marine litter (focusing on plastic), with at least eleven countries following to develop their own.[15]

In 2017, a regulation on plastics was felt around the world when the import ban China implemented on scrap plastics for recycling hit many of us at our kitchen recycle bin. Globally half of all scrap plastics used to be exported to China, and when China stopped taking them, the effects were felt in many locations—high-income countries in the Organization for Economic Cooperation and Development (OECD) were shipping plastic scrap to lower-income, non-OECD countries. Countries in South and Southeast Asia and Eastern Europe thought they were now the market for these materials. However, some countries, like the Philippines and Malaysia, refused or shipped the materials back.[16] In some cases, countries still take plastic scrap, but it may get burned and illegally dumped, like the example of UK's plastic scrap in Turkey.[17] The practice of exporting plastic scrap led to one of the fastest treaty amendments ever implemented. The Basel Convention controls international shipments of hazardous waste and waste destined for recycling or disposal (e.g., electronic waste). As of January 1, 2021, the treaty was amended to include most plastic scrap transboundary movements, requiring prior notice to and consent from the receiving country. Not all countries have ratified the Basel Convention, though, including the USA.

Based upon parallel studies in Science Magazine (2020), we know that "downstream" policies and interventions are not enough, the quantity of plastics entering the ocean is set to triple by 2040 in a business-as-usual scenario.[18] For example, 500,000 people per day would need to connect to waste collection alone to close the collection gap by 2040—this is impossible. Even with all the government and company policy commitments in place as of 2020, there would only be a 7% reduction of plastics entering our ocean. No one approach can do it, but upstream interventions, including reduction, reuse, and redesign, cannot be ignored while only focusing on collection, recycling, disposal, and leakage.

The public has begun realizing this as well and demanding better packaging and information from companies. Several lawsuits have shown that making claims and promises about goals to reduce plastic pollution, especially by simply increasing recycling, are not enough. Even expanding recycling for the sake of being more circular has to provide accurate information to the public. Lawsuits saying recyclability claims have been overstated or are not transparent have successfully gotten settlements against Dr. Pepper-Keurig and Terracycle. These resulted in label changes by Dr. Pepper-Keurig on their K-cups, Terracycle, and eight companies about recyclability claims: Coca-Cola, Procter & Gamble, Late July Snacks, Gerber, L'Oreal, Tom's of Maine, Clorox, and Materne. These suits helped garner support for and pass the California Truth-In-Labeling Law (SB343) in 2021. But California has gone further in getting to the root of spurious claims. In April 2022, in a first-of-its-kind investigation, the Attorney General of California announced an investigation into the fossil fuel and petrochemical industries for their role in causing and exacerbating the global plastics pollution crisis.[19] Specifically, the investigation will target companies that have caused and exacerbated the global plastics pollution crisis, investigate these companies' role in perpetuating myths around recycling and the extent to which this deception is still ongoing, and determine if these actions violate the law. Attorney General Bonta subpoenaed ExxonMobil seeking information relating to the case.[20] Both governments and the public are tired of carrying the burden of plastic waste management and pollution.

The worldwide patchwork of plastic policies has influenced the calls for a global agreement. A historic moment at the United Nations Environment Assembly (UNEA) 5.2 in March 2022 was witnessed around the world as the United Nations Environment Assembly broadcast the announcement of the plastic pollution resolutions live. It was titled the "Resolution to End Plastic Pollution: Towards an International Legally Binding Instrument." It was the first step for the member states to agree to create a legally binding agreement to address plastic pollution. The room erupted in applause. The numerous late nights of negotiations paid off as 193 countries agreed upon this resolution.

The UNEA plastics resolution builds on the foundation of other global initiatives. For example, in 2015, when Germany held the presidency of the G7 (Canada, France, Germany, Italy, Japan, UK, USA), the countries passed an action plan to combat marine litter. This was a first for the seven countries to agree to address this issue together. The action plan was for "Marine Litter" and not yet even defined plastic pollution. The first-of-its-kind G7 action plan included calls for international assistance, sharing best practices, and utilizing existing platforms like the Global Partnership on Marine Litter (GPML), created in 2012, promoting individual and corporate behavior change recognizing prevention is key, and supporting a broad range of available policies and instruments.

The UNEA resolution covers the entire life cycle of plastics from production to use, to the management of waste, and, finally, leakage. It mentions waste reduction as one component. It also mentions recycling in the context of a circular economy, keeping materials in the economy for as long as possible. This goal is broader than just recycling; it means upstream changes before recycling so that it would reduce waste generation in the first place. It then takes into account the design of products and materials for circular management at the end of their cycle. The resolution is also inclusive to the global informal collection and recycling community, which currently provides significant materials management and recycling worldwide, but often goes unrecognized while working in conditions that are not protective of their health and environment.

The reduction and prevention of plastic pollution go beyond the marine environment in the resolution; it applies to all environments. The UNEA resolution declares that policies and activities should be informed by science, including socioeconomic information as well. The resolution also allows for the flexibility for the future legally binding agreement to be applied in the national context of each member state—recognizing that there are differences between them. For example, what actions might make sense in one state, like a ban on single-use water containers, may not make sense in another. Resources for the member states were mentioned so that the burden does not entirely fall on the countries or local communities for implementation. A comprehensive framework seems to be in the resolution, but both assessing the effectiveness and progress of it require standardized data collection and monitoring, which is not happening at the required spatial or temporal scales as of 2022. As described by Inger Andersen, executive director of UNEP after UNEA, "Today marks a triumph by planet Earth over single-use plastics. This global agreement is the most significant environmental multilateral deal since the Paris accord. It is an insurance policy for this generation and future ones, so they may live with plastic and not be doomed by it."[21] We hope that by the time you read this, we have a legally binding framework for countries around the world to sign on to, that will finally reduce plastics entering our land and ocean, protecting human health and our environment.

9

ALTERNATIVES AND INTERVENTIONS FOR PLASTIC

In the previous chapters of this book, we covered the rapid increase in plastic production and environmental pollution. We discussed the implications of plastic pollution on the environment and society. And at this point, you might be wondering what can be done to remediate or "solve" this problem. Many people have thought the same thing. Over the decades, multiple approaches have been utilized to address plastic production, use, and waste management, including policy interventions, industry innovations, and community organizing (Table 1). Policy interventions are when local, national, and global governments implement guidance, laws, and regulations to reduce plastic production, use, or disposal. Many of these policies were discussed in chapter 8. Industry interventions include making alternate/reformulated/recycled materials, new waste management solutions, or buy back /reuse/refill programs. One example is Recyclebank, an industry-based rewards system for residents making green choices like recycling. Communities and nonprofit groups often organize people, raise resources and encourage collective action to address the plastic pollution crisis. To fully conceptualize all the work currently being done to address plastic pollution from production to leakage, it is best to walk through some commonly employed intervention methods.

Table 1 A list of interventions used to address plastic pollution around the world.

Intervention	Explanation	Example
Policy (chapter 8)	Local, national, and global governments put in place policies to regulate or reduce plastic production, use, or disposal.	Plastic bag bans/taxes Microbead bans Recycling programs
Industry	Industry is driven by market demand to create alternatives and solutions to plastic pollution.	Product buyback programs Bioplastics Biodegradable plastics Recycling innovation
Community/ Nonprofits	Consumers are organizing to drive market demand, policy, and cleaning up their communities.	Wastepickers Community cleanups Reduction of plastic use Refill/reuse schemes

Cleanup

Because the impacts on the environment and animals are so evident, our first instinct to address the plastic pollution crisis is to clean up the plastic and remove the debris from the environment. In the 1950s, beverage corporations and tobacco companies founded the non-profit Keep America Beautiful (KAB) to "promote a clean environment." However, as referenced in the introduction to this book, a significant controversy over their true motivations ensued when they refused to promote beverage bottle return schemes, which are proven to reduce bottle litter in the environment, but were not supported by their corporate funders. KAB started its "People start pollution. People can stop it," campaign in 1971 with an incredibly greenwashed and inappropriate ad, "the crying Indian." This ad placed the blame for litter squarely on the shoulders of the public while putting the industry in a position only to offer the public disposable items that maximize their profits.[1] KAB then asks the public to volunteer to clean up litter, including plastics, in terrestrial environments. The industry provided funding for work that KAB would do to reduce waste, promote recycling and "beautify" neighborhoods, primarily with volunteers. Although KAB affiliates often have locally positive impacts, the funding was ultimately self-serving, allowing the industry to continue in a business-as-usual setting, even increasing the quantities of disposable items making it into the marketplace. Other than more recently publishing periodic litter reports, which have now been used in scientific studies, there was previously little data collected to illustrate what was being found in the environment by KAB.

One of the first coastal cleanups to log data on each item collected was held in Texas in 1986. This annual one-day global cleanup and data collection event began to be known as the International Coastal Cleanup (ICC), organized by Ocean Conservancy for over three decades. As time went by, *thousands* of nonprofits and community organizations began to implement massive and small cleanups around the globe, sometimes in collaboration with scientists (academic and community based). But most of these cleanups require large numbers of local volunteers to manually pick up and sort trash to be recycled or landfilled. Both community members and scientists work to characterize the garbage collected to help with *the upstream* implementation of more interventions and alternatives to plastic waste. One such study from the organization Break Free from Plastic identifies the top company products that are most littered around the world, which in 2021 are, by rank,. 1) Coca-Cola, 2) PepsiCo, 3) Unilever, 4) Nestle, and 5) Proctor and Gamble.[2] This activism has led to some commitments from these companies to address the plastic pollution crisis; however, very little to none of these commitments include the reduction of producing single-use plastics. For example, in 2018, Coca-Cola created the World Without Waste initiative promising to collect and recycle the equivalent of a bottle or can for every one they sell by 2030, to make 100% of packaging recyclable by 2025, and to use 50% recycled material in bottles and cans by 2030. On-the-ground activism has begun to hold corporations accountable for the entire life cycle of their product, and data has given some power to the people. However, the downfall of utilizing only cleanups as an intervention for change is that it takes massive amounts of time and effort from volunteers who should not carry the burden.

Initially, the main focus was on the cleanup of litter on land, even along the coast, and less from fresh and marine waters. This shifted with the increasing images of plastics causing harm to marine animals through entanglement, accumulation of waste in their gut, and strangulation. Organizations also implemented more common strategies from land-based cleanups in water systems, including capturing floating debris on the surface. River cleanups are less often initiated than coastal and terrestrial litter cleanups, but freshwater streams and rivers are both a sink and a source of plastics bound for the sea. "Trash traps" are booms and containers that float in fresh waterways to capture the debris that floats on the surface. The deployment of trapping technology has prevented waste from flowing downstream that could have been bound for the ocean. But the caveat of this technology is that it requires infrastructure, operation, and maintenance to regularly remove the waste from the trash trap before the trash begins to accumulate, settle into the sediment, release chemicals into the water, or fragment into microplastics. These devices also need to be placed in a location with access to waste management infrastructure after collecting the plastic. If there is no waste management infrastructure on land to manage the waste collected, the pollution is just getting transferred from water to the ground—with the potential to get back into the water again.

Additionally, automated technologies have emerged, including autonomous robots that can pick up trash and solar- or water-powered stationary systems that skim surface water trash and deposit it directly into a dumpster. Beach sweepers are driven on the beach to comb through the shorelines to pick up more significant plastic trash. But, very few technologies exist to clean up microplastics in the environment. The Hula One is a vacuum that relies on the buoyancy of plastics to remove microplastics from sand, but it's time-intensive and not very efficient. Water treatment facilities do remove a majority of microplastics from wastewater, but then they end up in the sludge that remains. In the case of nanoplastics, there is currently no feasible technology to remove them from the environment.

Cleaning plastics from any water source is more complicated than in terrestrial environments because not all the plastic will stay on the surface. Some polymers float on the surface, some immediately sink, and others will sink after weathering or becoming small enough to be dropped by a changing density from the size or microbial growth. Suggestions to clean the oceans at the surface and other depths have not been practical. The sea is very, very deep. So deep that we have yet to map the world's seafloor fully. Yes, plastic has likely seen more than humankind has seen of the ocean. Cleanups are a very downstream solution to plastic pollution that puts the onus on consumers and volunteers to fix instead of the producers that make the products to sell. It is like trying to mop up the bathroom floor before you turn off the tap of an overflowing bathtub.

Change the Material and Design

So can we use a different material than plastic and redesign our packaging? The most straightforward idea is to return to the natural materials we started with, like metals, glass, and wood, to replace plastic. However, plastic has filled a vital need in various industries that these materials would not be able to efficiently or effectively replace, including medical and research supplies, consumer products, construction, and transportation. While medical and research supplies have minimal alternatives for plastic materials, the packaging industry and consumer products are essential for developing other options and solutions.

As discussed in chapter 6, we have seen extensive imagery of wildlife being strangled and entangled in six-pack rings. An extensive public information campaign made the public aware of the problem and encouraged them to cut up the plastic rings that had choked so many seals. Because of this campaign, millions of people still cut up the six-pack rings to this day. The harm of six-pack rings to wildlife led to a regulation that required the rings to fragment in the environment. However, in many cases, six-pack rings are a thing of the past. As a redesign, cans began to be wrapped entirely in a sturdy film plastic or just come in paperboard boxes, which used to be just for twelve-packs but now can be for six- or eight-packs of beverages. There are now rigid plastic can holders that can be reused, and even just adhesive that holds the cans together, reducing material and plastic use by 76% (adhesives are plastic).[3] There is even a biodegradable six-pack ring that, at its origin, was made from barley and wheat remnants from the brewing process, but now it is made from "fiber," which was touted as fish food (we don't recommend this!). Still, as a fiber product, similar to paper, it should biodegrade in a landfill or compost system and in the environment.

The six-pack ring is an example of redesign and material substitution. However, there may be tradeoffs to some of these choices, including energy and water use and carbon emissions for various points along production and distribution. However, avoiding the extraction, processing, and refinement of fossil fuels, as well as protecting the environment from plastic pollution, is often enough motivation to make these changes—and there is not a very good way to compare these tradeoffs to each other.

Another successful redesign example also had to do with aluminum beverage cans. Aluminum cans used to have a tab that had to be entirely removed to get it open to consume the beverage. But these tabs we found in the environment and could easily cut someone if stepped on because of their shape and sharpness of the material. The tabs were redesigned to open the can but remain on the can as we see today, allowing cans to remain the packaging of choice for many beverages. Although once polyethylene terephthalate (PET) was invented and was strong enough to hold carbonated liquids, as well as the membrane in the cap to keep the carbonation in, plastic took over the market share in many cases. Now, plastic caps are more often found than bottles in the environment, begging the question of innovations to alter bottle design to keep them attached to the bottle, something Coca-Cola is now implementing by 2024 in Great Britain.

While many materials scientists are working hard on materials that serve similar purposes to plastics but are not made of fossil fuels or don't have persistence in the environment, they have yet to be a large part of the market. Compostable plastics, like polylactic acid (PLA), will not biodegrade if mismanaged but are not made of fossil fuels and will compost in an industrial setting (not your home compost). In other cases, there are now polymers that will biodegrade in soils and your home compost, like Polyhydroxyalkanoate (PHA). However, biodegradability in ocean environments is still challenging since the microbial activity there can be limited or significantly different from the microbes in soils. Petroleum-based plastics that are not recyclable, like thin-film plastics used to make food wrappers, could be a logical application for biodegradable plastics.

If designing for recycling, in an ideal world, all plastics would be standardized with just a few primary polymer types with minimal colors and chemical additives that are nontoxic. Layered and hybrid materials that can't be recycled would be restricted. If we had followed Green Engineering Principles like this in the first place, we could have avoided many of the consequences that we now see.[4] For example, the material should be able to be easily recycled multiple times, shouldn't fall apart into microplastics, shouldn't contain harmful chemicals, should be easily recycled with other plastics, and should be used for a specific need and not just for the sake of using raw materials to make plastic. If needed, using plastic as a material should be done with intention, taking an "output pulled" versus "input pushed" approach, limiting what is made from plastics.

There are essential drivers to some plastic use. The best example comes from a plastic straw redesign campaign that tried to make straws of metal or paper. The disability activism community immediately educated others that some people can only have plastic straws. In some cases, reusable silicone straws may work, but straws are now one of the first use cases for PHA, which have the flexibility and mouth feel of plastic but are biodegradable. In addition, large facilities, including major movie theater chains, wanted to move to alternative material straws. In terms of redesign, major companies like PepsiCo and Unilever have committed to making all their packaging recyclable or compostable, requiring both product redesign and potentially compostable or biodegradable materials. The key to all of these commitments is the definitions of "recyclable" and "biodegradable" versus "compostable"—all of which have to be matched with the materials management infrastructure geographically. Other redesign examples include to-go containers that are not expanded polystyrene (EPS) and party cups that are aluminum and not plastic.[5] In some coastal areas, a certification for "ocean-friendly" restaurants has helped establishments in communities reduce the use of plastic and procure alternative materials, like reusable to-go container products. But alternative materials are often more expensive, and not all businesses or people can afford them, nor does everyone have access to reuse schemes or choices in materials at the store.

One major drawback of alternative materials and redesign is that it could continue the disposable norms that we currently have, just with different materials. And there is even a possibility that switching to materials that people perceive as more "environmentally friendly" and not an issue for plastic pollution means they feel free to use more than they would otherwise, resulting in overconsumption of new material instead of reducing the use of an unnecessary product. This is known as Javon's paradox, which found that people would drive more with an electric car that they felt was better for the environment while they avoided driving or combined trips, etc., with a gas car. An example with plastic is if people avoid getting plastic utensils by bringing a reusable set with them, they may then stop doing that when the utensils become nonplastic or biodegradable.

Circular Economy

The subsequent intervention commonly suggested to deal with the accumulation of plastic waste is to better manage it in ways that prevent it from ending up in the environment. This concept is commonly associated with creating a circular economy and is related to the previous section on alternative materials and redesign since a fully circular economy would require both. A circular economy is an idea of avoiding waste creation and then sustainably recirculating goods produced and consumed via sharing, reusing, refurbishing, and recycling. If properly implemented, it would equate to zero waste in communities. It could also indicate the decline in the need for raw materials to make new products because we would be reusing the products already in circulation. This concept has been popularized in a variety of ways. The first is via the influence of the media. Social media influencers are known to post about their zero waste lifestyles or new thrifting hauls. The Disney live-action Cruella movie displays the idea of reusing fabrics for fashion. However, the plastic pollution portrayal in Hollywood media needs to be improved based on a study by the USC Annenberg Norman Lear Center.[6] They found that single-use plastics were common on TV, appearing in every single episode, with an average of 28 single-use plastic items per episode, 93% of which were not disposed of on-screen, lending to the "waste just disappears" mentality. When waste disposal was shown, a whopping 80% was shown to be littered, often the most common items found in our environment, like Carrie Bradshaw's cigarette in "And Just Like That . . ." Media should not illustrate that it is stylish or acceptable to litter anything.

Companies are also embracing this idea. For example, Starbucks coffee offers reusable options made from recycled materials and encourages patrons to bring their reusable mugs. But a deeper case study can be seen in Ikea. Ikea is often purchased as furniture for a college experience where it is used for relatively short periods and sometimes even within just a few months of summer. At the end of every dorm move-out, Ikea furniture can be seen haphazardly accumulating in trash receptacles. Ikea is known to be affordable and potentially disposable since it is reported to be easy to put together, take apart, and dispose. Ikea recognized the short-term nature of their furniture compared to other brands and saw this as an opportunity to keep the furniture out of the environment and their furniture in circulation by creating a buy back and resell program.[7]

We have briefly discussed some of the difficulties with waste management in previous chapters, including chemical and mechanical recycling. While there have been recent goals set to further develop the technology for advanced chemical and mechanical recycling, plastics are not like other materials, like metals, that can be infinitely recycled. Virgin materials are still needed to make a product with similar integrity as the original product. The other issue with recycling lies in the types of products that can be recycled which is relatively limited by plastic type and the size of the plastic as nothing smaller than a deck of cards will be recycled by waste management systems. Some industry producers have stepped up to address the recyclability issue of their plastics by changing the material design (sprite bottles are now clear instead of green) and creating specific collection systems to recycle their products. In addition, some companies accept back their products to be recycled, including K-cups, Taco Bell sauce packets, and Biotrue contact lenses; however, the onerous duty is on the public to return those items to drop-off centers or ship them back.

One of the more significant issues arising from creating a truly circular economy lies in the packaging industry, which uses a large proportion of plastic produced to transport goods globally. Most often, the packaging, other than cardboard boxes, is used once and is not often recycled. Some industries have worked to grant a second life use for these products in items for infrastructure, like using single-use plastic bags to make bricks in Kenya and outdoor patio decks. However, it doesn't reinforce a circular economy because it requires those bags to be replaced, so their source of the material should be avoided or come from a recycled source as well. Some industries like Amazon have tried new packaging envelopes using smaller quantities of plastic, but these aren't recyclable everywhere.

In 2022, a Department of Energy report and a report by The Last Beach Cleanup and Beyond Plastics both showed that the US recycling rate dropped from 8.7% in 2018 to 5–6%.[8] We still still confused about which plastics to recycle wherever we are, when in reality, the Beyond Plastics report says the material itself can't be realistically recycled because of the many challenges along the way also outlined in chapter 3.[9] A 2022 analysis conducted by Eunomia Research & Consulting found that if the top five beverage companies meet their pledges on recycling (which remains to be seen), these pledges would only reduce aquatic pollution from single-use plastic bottles by 7%. While plastic materials management and recycling will be a part of the zero-waste future, they will not even come close to solving the plastic pollution crisis alone.

Full Stop: Plastic Avoidance

And while the first three solutions could and are being implemented today, they still have yet to bear fruit in reducing the amount of plastic waste produced or emitted into the environment. At this point, recycling has not reduced virgin production. And all the commitments made to date only project a growing quantity of plastic entering our environment and waterways in the next ten to thirty years.

The most thought-provoking solution is to stop the use of plastic entirely. Now, what would that look like? Does that mean we end capitalism and consumerism as we know it today? Do we go back to the 1940s ceramic pots and hand-woven cotton clothing? Do we abandon our life-saving medical supplies and transportation options? NGOs and some scientists believe that the only way we reduce plastic pollution is to place a cap on plastic production. With most plastics used for packaging, at least removing as much of this as possible could be a start (see the unnecessary, avoidable, and problematic plastics section in chapter 2). Durable and life-saving plastics might be another story. How would we meet people's needs and get products to them without plastics? This concept is currently being piloted within cities by an organization called Perpetual Use—overcoming the burdens and exploring how to scale reuse to be inclusive and accessible. Imagine walking into a town where every packaging item, whether plastic or not, is continuously reused.

What Can We Do?

The most effective method to address the issue of plastic pollution is to have complete systems-based solutions that require technologies, company investment, policy changes, and population-driven changes. This includes first determining products/polymers that are unnecessary, avoidable, and problematic. Once these polymeric materials and polymer-associated chemicals are determined, they should be locally or globally banned in the marketplace. A product ban will also spur the demand for creating or utilizing alternative materials and products. However, to make these alternatives competitive, governments must remove incentives for using fossil fuels for nondurable plastics to give an alternative feedstock a chance. Second, we must develop infrastructure and policies to support a reuse system of

Imagine walking into a town where every packaging item, whether plastic or not, is continuously reused.

delivery of food and products that don't produce waste in the first place. And finally, we need more robust waste infrastructure development for collection, disposal, recycling, composting, and environmental cleanup of garbage. All of these choices must be made collectively in locations where plastic production, use, and disposal are significant.

In addition to systemic changes, individual actions have an essential effect on driving change. So what can we do? Here are a few ideas. First, you can make feasible decisions for your lifestyle via how you spend your money, how you use your voice, and how you get involved in government systems (dollar, voice, and vote). If you are concerned about your health or the environmental impacts of plastic, you can make decisions to limit unnecessary plastic exposure. You can do this by purchasing reusable or refillable plastic-free items. Suppose you want companies to create more reusable, biodegradable, or fewer plastic-containing options. In that case, you can use your voice by talking with people about plastic pollution, writing emails to industries asking for change, or working with local businesses to encourage more reusable or biodegradable take-out options. You can also work with local governments to request improved recycling and compost facilities that accept biodegradable/compostable plastics. Start a community garden or farmers market to avoid packaged products. Join a local cleanup and provide that data to Debris Tracker and international cleanup campaigns. Talk to everyone you know about this issue or send them this book and we can all change our mindset on plastic. You can make many small changes to make this plastic pollution issue heard and addressed.

The alternatives and solutions for the plastic pollution issue are just as complicated as the thousands of formulations and forms plastic can take on. So while there is not one simple solution, there is a vast community of eager people working to address the issues created by plastic production and pollution.

GLOSSARY

Polymer
Also known as plastic. Polymers are made of up individual monomeric compounds that are linked together to form a chain. Polymers are used to make a variety of plastic products and have many different physical chemical properties and functions that are dependent on the monomer used.

Polymer-associated chemicals
Sometimes referred to as plastic additives. These are chemicals incorporated into plastic products intentionally to fulfill a specific function or aesthetic for the material or unintentionally due to the breakdown of other chemicals during processing or contamination.

Bisphenol A (BPA)
A polymer-associated chemical used as either a monomer or plastic additive (antioxidant) in some plastics. This chemical is found to be released from plastics into the human body and environment. BPA is also determined to be an endocrine-disrupting chemical and linked to some forms of cancer.

Environmental justice
The equitable treatment and involvement of all people with respect to the development, implementation, and enforcement of environmental laws, regulations and policies.

Entanglement
For an organism to become wrapped and twisted into a mass sometimes by marine litter and debris including plastic fishing line, balloons, or plastic bags.

Bioplastic
Plastics derived from natural polymers found in organisms with plastic-like properties.

Synthetic plastic
Plastics that use fossil fuels as the feedstock to make their materials.

Biodegradable plastic

Plastics that can be decomposed by the action of living organisms, such as microbes into water, carbon dioxide, and biomass. These plastics can be bioplastics, synthetic plastics, or a combination of the two. Some degradation also requires specific conditions such as compostable facilities that provide high heat, varied oxygen levels, and pressure to accelerate the process.

Planetary boundary threat

Human-derived perturbations of earth systems that do not fit within the bounds of the environmental boundaries of the planet come at risk of causing serious and abrupt environmental changes. An example of this can be seen in climate change.

Microplastic

Microplastics are plastics smaller than 5 millimeters in size but larger than 1,000 micrometers. Microplastics can be produced to be these size ranges (primary) or form from degradation of macrosized plastics (secondary). Plastics smaller than this size range are defined to be nanoplastics.

Single-use plastic

Plastic materials designed to be used only once and then discarded. Examples are plastic cups, utensils, food wrappers, plastic bottles, etc.

Great Pacific Garbage Patch

A misnomer, it is not an "island" in the Pacific, but the accumulation of plastic located in the Great Pacific Gyre (the circulation of currents in the northern Pacific Ocean). The accumulation does not look like an island. Some plastic objects may be observed, but much of the plastic is in broken up pieces and is so microscopic (microplastics) that it can't be seen until it is pulled out of the water with a sampling net.

Municipal solid waste

This is waste that community members generate at the household level. It includes waste generated at the workplace and at restaurants as well. It does not include waste generated from construction, demolition, manufacturing, or industrial processes.

Litter
Waste that is uncontrolled in the environment. Land-based litter is often called "litter"—while litter in the marine environment is called "marine litter." It does not imply responsibility.

Marine debris
The US National Oceanic and Atmospheric Administration (NOAA) definition of marine debris is anything that does not naturally belong in the water. It includes plastics and lost or abandoned fishing gear.

Circular economy
According to the Save Our Seas 2.0 Act in the United States, the circular economy refers to an economy that uses a systems-focused approach and involves industrial processes and economic activities that are restorative or regenerative by design, enable resources used in such processes and activities to maintain their highest value for as long as possible, and aim for the elimination of waste through the superior design of materials, products, and systems (including business models).

Plastic bans/fees
Plastic bans prohibit a particular type or form of plastic from being sold or used in a certain jurisdiction. Plastic fees impose a fee for use or tax when a particular type or form of plastic from being sold or used in a certain jurisdiction.

Deposit/return scheme
People who purchase an item pay a small deposit which can be refunded upon return of the packaging to a convenient location, like a store or redemption center. They are most often used for glass or plastic beverage bottles.

NOTES

Chapter 1

1. https://www.merriam-webster.com/dictionary/plastic.
2. Susan Freinkel, *Plastic: A Toxic Love Story* (Boston: Mariner Books, 2011).
3. Laura Sullivan, "How Big Oil Misled the Public into Believing Plastic Would Be Recycled," National Public Radio, September 11, 2020, https://www.npr.org/2020/09/11/897692090/how-big-oil-misled-the-public-into-believing-plastic-would-be-recycled.

Chapter 2

1. Roland Geyer, Jenna R. Jambeck, and Kara Lavender Law, "Production, Use, and Fate of All Plastics Ever Made," *Science Advances* 3, no. 7 (2017): e1700782.
2. Plastics Europe, *Plastics—the Facts 2020*, September 2021, https://plasticseurope.org/knowledge-hub/plastics-the-facts-2020/.
3. Plastics Europe, *Plastics—the Facts 2020*.
4. Data only recorded in the EU27+3 (UK, Norway, and Switzerland) by Plastics Europe. Plastics Europe, *Plastics—the Facts 2020*.
5. United States Environmental Protection Agency, "Advancing Sustainable Materials Management: 2017 Fact Sheet Assessing Trends in Material Generation, Recycling, Composting, Combustion with Energy Recovery and Landfilling in the United States," Washington, DC, 2020.
6. Ocean Conservancy, *The Beach and Beyond: International Coastal Cleanup 2019 Report*, September 2019, https://oceanconservancy.org/wp-content/uploads/2019/09/Final-2019-ICC-Report.pdf; Ocean Conservancy, *Together, We Are Team Ocean—2020 Report*, October 2020, https://oceanconservancy.org/wp-content/uploads/2020/10/FINAL_2020ICC_Report.pdf; Kathryn Youngblood, Sheridan Finder, and Jenna Jambeck, *Mississippi River Plastic Pollution Initiative 2021 Science Report*, September 2021, https://www.unep.org/resources/report/mississippi-river-plastic-pollution-initiative-2021-science-report.
7. Australia Packaging Covenant Organization (APCO), *Single-Use, Problematic and Unnecessary Plastic Packaging*, October 2020, https://documents.packagingcovenant.org.au/public-documents/Single-Use%20Problematic%20and%20Unnecessary%20Plastic%20Packaging; U.S. Plastics Pact, *Problematic and Unnecessary Materials Report*, January 2022, https://usplasticspact.org/wp-content/uploads/dlm_uploads/2022/01/U.S.-Plastics-Pact-Problematic-Unnecessary-Materials-Report-1.25.2022.pdf.

Chapter 3

1. Silpa Kaza et al., *What a Waste 2.0: A Global Snapshot of Solid Waste Management to 2050* (Washington, DC: World Bank Group, 2018), https://openknowledge.worldbank.org/handle/10986/30317.
2. Roland Geyer, Jenna R. Jambeck, and Kara Lavender Law, "Production, Use, and Fate of All Plastics Ever Made," *Science Advances* 3, no. 7 (2017): e1700782.
3. Geyer, Jambeck, and Law, "Production, Use, and Fate."
4. Eileen Maura McGurty, "Warren County, NC, and the Emergence of the Environmental Justice Movement: Unlikely Coalitions and Shared Meanings in Local Collective Action," *Society & Natural Resources* 13, no. 4 (2000): 373–387; Robert D. Bullard, *Dumping in Dixie: Race, Class, and Environmental Quality* (Boulder, CO: Westview, 1990); Clare Cannon, "Examining Rural Environmental Injustice: An Analysis of Ruralness, Class, Race, and Gender on the Presence of Landfills across the United States," *Journal of Rural and Community Development* 15, no. 1 (2020): 89–114; Omolade Erogunaiye, "Environmental Justice and Landfills: Application of Spatial and Statistical Analysis for Assessing Landfill Sites in Texas," (PhD diss., Texas Southern University, 2019).
5. Kenneth D. Tunnell, "Illegal Dumping: Large and Small Scale Littering in Rural Kentucky," *Journal of Rural Social Sciences* 23, no. 2 (2008): 29–42.
6. Geyer, Jambeck, and Law, "Production, Use, and Fate."
7. Plastics Waste Management Institute, *An Introduction to Plastic Recycling 2019*, January 2020, https://www.pwmi.or.jp/ei/plastic_recycling_2019.pdf.
8. Amy L. Brooks, Shunli Wang, and Jenna R. Jambeck, "The Chinese Import Ban and Its Impact on Global Plastic Waste Trade," *Science Advances* 4, no. 6 (2018): eaat0131.
9. Kara Lavender Law et al., "The United States' Contribution of Plastic Waste to Land and Ocean," *Science Advances* 6, no. 44 (2020): eabd0288.
10. Geyer, Jambeck, and Law, "Production, Use, and Fate."
11. Pinjing He et al., "Municipal Solid Waste (MSW) Landfill: A Source of Microplastics? Evidence of Microplastics in Landfill Leachate," *Water Research* 159 (2019): 38–45.
12. Geyer, Jambeck, and Law, "Production, Use, and Fate."
13. Christine Wiedinmyer, Robert J. Yokelson, and Brian K. Gullett, "Global Emissions of Trace Gases, Particulate Matter, and Hazardous Air Pollutants from Open Burning of Domestic Waste," *Environmental Science & Technology* 48, no. 16 (2014): 9523–9530; Brian K. Gullett et al., "Emissions of PCDD/F from Uncontrolled, Domestic Waste Burning," *Chemosphere* 43, no. 4 (2001): 721–725; Costas A. Velis and Ed Cook, "Mismanagement of Plastic Waste

through Open Burning with Emphasis on the Global South: A Systematic Review of Risks to Occupational and Public Health," *Environmental Science & Technology* 55, no. 11 (2021): 7186–7207.

Chapter 4

1. Ocean Conservancy, "Fighting for Trash Free Seas: International Coastal Cleanup," accessed May 2022, https://oceanconservancy.org/trash-free-seas/international-coastal-cleanup/.
2. Laurent Lebreton et al., "Evidence that the Great Pacific Garbage Patch Is Rapidly Accumulating Plastic," *Scientific Reports* 8, no. 1 (2018): 1–15.
3. Kara Lavender Law et al., "Plastic Accumulation in the North Atlantic Subtropical Gyre," *Science* 329, no. 5996 (2010): 1185–1188.
4. Marcus Eriksen et al., "Plastic Pollution in the World's Oceans: More Than 5 Trillion Plastic Pieces Weighing over 250,000 Tons Afloat at Sea," *PLoS ONE* 9, no. 12 (2014): e111913.
5. Steve Allen et al., "Atmospheric Transport and Deposition of Microplastics in a Remote Mountain Catchment," *Nature Geoscience* 12, no. 5 (2019): 339–344.
6. X. Peng et al., "Microplastics Contaminate the Deepest Part of the World's Ocean," *Geochemical Perspectives Letters* 9, no. 1 (2018): 1–5.
7. Oisín Ó. Briain et al., "The Role of Wet Wipes and Sanitary Towels as a Source of White Microplastic Fibres in the Marine Environment," *Water Research* 182 (2020): 116021.
8. Steve Allen et al., "Examination of the Ocean as a Source for Atmospheric Microplastics," *PLoS ONE* 15, no. 5 (2020): e0232746.
9. Janice Brahney et al., "Plastic Rain in Protected Areas of the United States," *Science* 368, no. 6496 (2020): 1257–1260.

Chapter 5

1. Hans Zwiefel, Ralph D. Meier, and Michael Schiller, Plastics Additives Handbook, 6th ed. (Cincinnati, OH: Hanser Publications, 2009).
2. Sam Kean, "When Plastics Are Precious," *Science* 373, no. 6550 (2021): 40–42.
3. Helene, Wiesinger, Zhanyun Wang, and Stefanie Hellweg, "Deep Dive into Plastic Monomers, Additives, and Processing Aids," *Environmental Science & Technology* 55, no. 13 (2021): 9339–9351.
4. Rudolph D. Deanin, "Additives in Plastics," *Environmental Health Perspectives* 11 (1975): 35–39.
5. Jaromir Michałowicz, "Bisphenol A–Sources, Toxicity and Biotransformation," *Environmental Toxicology and Pharmacology* 37, no. 2 (2014): 738–758.

6. Ksenia J. Groh et al., "Overview of Known Plastic Packaging-Associated Chemicals and Their Hazards," *Science of the Total Environment* 651 (2019): 3253–3268.
7. John N. Hahladakis et al., "An Overview of Chemical Additives Present in Plastics: Migration, Release, Fate and Environmental Impact During Their Use, Disposal and Recycling," *Journal of Hazardous Materials* 344 (2018): 179–199.
8. Emma L. Teuten et al., "Transport and Release of Chemicals from Plastics to the Environment and to Wildlife," *Philosophical Transactions of the Royal Society B: Biological Sciences* 364, no. 1526 (2009): 2027–2045.

Chapter 6

1. Matthew MacLeod et al., "The Global Threat from Plastic Pollution," *Science* 373, no. 6550 (2021): 61–65.
2. Karl W. Kenyon and Eugene Kridler, "Laysan Albatrosses Swallow Indigestible Matter," *The Auk* 86, no. 2 (1969): 339–343.
3. Robson G. Santos, Gabriel E. Machovsky-Capuska, and Ryan Andrades, "Plastic Ingestion as an Evolutionary Trap: Toward a Holistic Understanding," *Science* 373, no. 6550 (2021): 56–60.
4. Marcus Eriksen et al., "The Plight of Camels Eating Plastic Waste," *Journal of Arid Environments* 185 (2021): 104374.
5. Associated Press, "Elephants Are Dying from Eating Plastic Waste in Sri Lankan Dump," New York Post, January 14, 2022, https://nypost.com/2022/01/14/elephants-dying-from-eating-plastic-waste-in-sri-lankan-dump/.
6. Roman Lehner et al., "Emergence of Nanoplastic in the Environment and Possible Impact on Human Health," *Environmental Science & Technology* 53, no. 4 (2019): 1748–1765.
7. Hans Zwiefel, Ralph D. Meier, and Michael Schiller, Plastics Additives Handbook, 6th ed. (Cincinnati, OH: Hanser Publications, 2009).
8. Charles Staples et al., "A Review of the Environmental Fate, Effects, and Exposures of Bisphenol A," *Chemosphere* 36, no. 10 (1998): 2149–2173.
9. Mark Anthony Browne et al., "Microplastic Moves Pollutants and Additives to Worms, Reducing Functions Linked to Health and Biodiversity," *Current Biology* 23, no. 23 (2013): 2388–2392.
10. Lauren Roman et al., "Plastic, Nutrition and Pollution; Relationships between Ingested Plastic and Metal Concentrations in the Livers of Two *Pachyptila* Seabirds," *Scientific Reports* 10, no. 1 (2020): 1–14.
11. Ksenia J. Groh et al., "Overview of Known Plastic Packaging-Associated Chemicals and Their Hazards," *Science of the Total Environment* 651 (2019): 3253–3268.

12. Lisa Zimmermann et al., "What Are the Drivers of Microplastic Toxicity? Comparing the Toxicity of Plastic Chemicals and Particles to *Daphnia magna*," *Environmental Pollution* 267 (2020): 115392.
13. Zhenyu Tian et al., "A Ubiquitous Tire Rubber-Derived Chemical Induces Acute Mortality in Coho Salmon," *Science* 371, no. 6525 (2021): 185–189.
14. Daisy Harley-Nyang et al., "Investigation and Analysis of Microplastics in Sewage Sludge and Biosolids: A Case Study from One Wastewater Treatment Works in the UK," *Science of the Total Environment* 823 (2022): 153735.
15. Valencia Beckwith and Mariana M. P. B. Fuentes, "Microplastic at Nesting Grounds Used by the Northern Gulf of Mexico Loggerhead Recovery Unit," *Marine Pollution Bulletin* 131 (2018): 32–37.
16. Helen V. Ford et al., "The Fundamental Links between Climate Change and Marine Plastic Pollution," *Science of the Total Environment* 806 (2022): 150392.

Chapter 7

1. Neelesh K. Varde and Daniel W. Pack, "Microspheres for Controlled Release Drug Delivery," *Expert Opinion on Biological Therapy* 4, no. 1 (2004): 35–51.
2. Sarah Gibbens, "You Eat Thousands of Bits of Plastic Every Year," *National Geographic*, June 5, 2019, https://www.nationalgeographic.com/environment/article/you-eat-thousands-of-bits-of-plastic-every-year.
3. Zehua Yan et al., "Analysis of Microplastics in Human Feces Reveals a Correlation between Fecal Microplastics and Inflammatory Bowel Disease Status," *Environmental Science & Technology* 56, no. 1 (2021): 414–421.
4. Sara B. Fournier et al., "Nanopolystyrene Translocation and Fetal Deposition after Acute Lung Exposure during Late-Stage Pregnancy," *Particle and Fibre Toxicology* 17, no. 1 (2020): 1–11.
5. Damian Carrington, "Microplastics Found in Human Blood for First Time," *The Guardian*, March 24, 2022.
6. Joshua Brockman, "Breathing Machine Recall Over Possible Cancer Risk Leaves Millions Scrambling for Substitutes," *New York Times*, August 17, 2021, https://www.nytimes.com/2021/08/17/health/cpap-breathing-devices-recall.html.
7. World Health Organization, *Dietary and Inhalation Exposure to Nano- and Microplastic Particles and Potential Implications for Human Health* (Geneva: World Health Organization, 2022).
8. Zahra Pilevar et al., "Migration of Styrene Monomer from Polystyrene Packaging Materials into Foods: Characterization and Safety Evaluation," *Trends in Food Science & Technology* 91 (2019): 248–261.

9. U.S. Plastics Pact, *Problematic and Unnecessary Materials Report*, January 2022, https://usplasticspact.org/wp-content/uploads/dlm_uploads/2022/01/U.S.-Plastics-Pact-Problematic-Unnecessary-Materials-Report-1.25.2022.pdf.
10. Danni Cui, Xuerong Li, and Natalia Quinete, "Occurrence, Fate, Sources and Toxicity of PFAS: What We Know So Far in Florida and Major Gaps," *Trends in Analytical Chemistry* 130 (2020): 115976.
11. Liza Gross, "Flame Retardants in Consumer Products Are Linked to Health and Cognitive Problems," *Washington Post*, April 15, 2013.
12. Johannes Völker et al., "Adipogenic Activity of Chemicals Used in Plastic Consumer Products," *Environmental Science & Technology* 56, no. 4 (2022): 2487–2496.
13. United Nations Environment Programme, *Neglected: Environmental Justice Impacts of Marine Litter and Plastic Pollution* (Nairobi: United Nations Environment Programme, 2021).
14. Center for International Environmental Law (CIEL) et al., *Plastic and Climate: The Hidden Costs of a Plastic Planet*, May 2019, https://www.ciel.org/plasticandclimate/.
15. Meg Parsons, Lara Taylor, and Roa Crease, "Indigenous Environmental Justice within Marine Ecosystems: A Systematic Review of the Literature on Indigenous Peoples' Involvement in Marine Governance and Management," *Sustainability* 13, no. 8 (2021): 4217; Max Liboiron, Rui Li, Elise Earl, and Imari Walker-Franklin, "Models of Justice Evoked in Published Scientific Studies of Plastic Pollution," *Facets*. 2023.
16. Max Liboiron et al., "Abundance and Types of Plastic Pollution in Surface Waters in the Eastern Arctic (Inuit Nunangat) and the Case for Reconciliation Science," *Science of the Total Environment* 782 (2021): 146809.

Chapter 8

1. Guinness World Records, "Largest Collection of Paper and Plastic Bags," accessed May 2022, https://www.guinnessworldrecords.com/world-records/68623-largest-collection-of-paper-and-plastic-bags.
2. Guinness World Records, "Most Ubiquitous Consumer Item," accessed May 2022, https://www.guinnessworldrecords.com/world-records/89873-most-ubiquitous-consumer-item.
3. United Nations Environment Program (UNEP), "From Birth to Ban: A History of the Plastic Shopping Bag," December 20, 2021, https://www.unep.org/news-and-stories/story/birth-ban-history-plastic-shopping-bag.
4. Guinness World Records, "First Documented Case of Plastic Entanglement," accessed May 2022, https://www.guinnessworldrecords.com/world-records/645219-first-documented-case-of-plastic-entanglement.

5. Jennifer Romer, "The Surfrider Foundation Releases Interactive Map of U.S. Plastic Reduction Policies," March 30, 2021, https://www.surfrider.org/coastal-blog/entry/the-surfrider-foundation-releases-interactive-map-of-u.s-plastic-reduction-policies.

6. Nicholas Institute for Environmental Policy Solutions, "Plastic Policy Inventory," Duke University, 2022, https://nicholasinstitute.duke.edu/plastics-policy-inventory.

7. Zoie Diana et al., "The Evolving Global Plastics Policy Landscape: An Inventory and Effectiveness Review," *Environmental Science & Policy* 134 (2022): 34–45.

8. Guinness World Records, "Most Ubiquitous Consumer Item."

9. Qamar Schuyler et al., "Economic Incentives Reduce Plastic Inputs to the Ocean," *Marine Policy* 96 (2018): 250–255.

10. United States Food and Drug Administration (FDA), "The Microbead-Free Waters Act: FAQs," February 25, 2022, https://www.fda.gov/cosmetics/cosmetics-laws-regulations/microbead-free-waters-act-faqs.

11. California Ocean Protection Council, *Statewide Microplastics Strategy: Understanding and Addressing Impacts to Protect Coastal and Ocean Health*, February 2022, https://www.opc.ca.gov/webmaster/ftp/pdf/agenda_items/20220223/Item_6_Exhibit_A_Statewide_Microplastics_Strategy.pdf.

12. Romer, "The Surfrider Foundation Releases Interactive Map."

13. Chelsea M. Rochman et al., "Classify Plastic Waste as Hazardous," *Nature* 494, no. 7436 (2013): 169–171.

14. Ephrat Livni, "Africa Is Leading the World in Plastic Bag Bans," Quartz Africa, May 18, 2019, https://qz.com/africa/1622547/africa-is-leading-the-world-in-plastic-bag-bans/; Laura Parker, "Plastic Bag Bans Are Spreading. But Are They Truly Effective?," *National Geographic*, April 17, 2019, https://www.nationalgeographic.com/environment/article/plastic-bag-bans-kenya-to-us-reduce-pollution.

15. Global Partnership on Marine Litter (GPML), "Action Plans," United Nations Environment Programme, accessed May 2022, https://www.gpmarinelitter.org/what-we-do/action-plans.

16. Reality Check Team, "Why Some Countries Are Shipping Back Plastic Waste," BBC News, June 2, 2019, https://www.bbc.com/news/world-48444874.

17. Kathryn Snowdon, "UK Plastic Waste Being Dumped and Burned in Turkey, Says Greenpeace," BBC News, May 17, 2021, https://www.bbc.com/news/uk-57139474

18. Pew Charitable Trusts and SYSTEMIQ, *Breaking the Plastic Wave: A Comprehensive Assessment of Pathways Towards Stopping Ocean Plastic Pollution*, 2020,

https://www.pewtrusts.org/-/media/assets/2020/10/breakingtheplasticwave_mainreport.pdf.
19. State of California Department of Justice, Office of the Attorney General, "Plastics," accessed May 2022, https://oag.ca.gov/plastics.
20. California Ocean Protection Council, *Statewide Microplastics Strategy*.
21. "Nations Sign Up to End Global Scourge of Plastic Pollution," UN News, March 2, 2022, https://news.un.org/en/story/2022/03/1113142.

Chapter 9

1. Ginger Strand, "The Crying Indian," *Orion Magazine* 20 (2008).
2. Break Free From Plastic, *Brand Audit Report 2021*, October 2021, https://www.breakfreefromplastic.org/wp-content/uploads/2021/10/BRAND-AUDIT-REPORT-2021.pdf.
3. Rachel Arthur, "Shrinking Away from Shrink Wrap: How Carlsberg Developed Its Six-Pack Glue Technology," Beverage Daily, September 28, 2021, https://www.beveragedaily.com/Article/2018/11/20/Shrinking-away-from-shrink-wrap-How-Carlsberg-developed-its-six-pack-glue-technology.
4. Paul T. Anastas and Julie B. Zimmerman, "Peer Reviewed: Design through the 12 Principles of Green Engineering," *Environmental Science & Technology* 37, no. 5 (2003): 94A–101A.
5. Jesse Klein, "Should You Swap Plastic for Aluminum Packaging? It's Complicated," Green Biz, March 19, 2021, https://www.greenbiz.com/article/should-you-swap-plastic-aluminum-packaging-its-complicated.
6. Plastic Pollution Coalition, "Flip the Script on Plastics," accessed May 2022, https://www.plasticpollutioncoalition.org/flipthescript.
7. Parija Kavilanz, "Ikea Will Pay You to Get Its Old Furniture Back," CNN Business, March 31, 2022, https://www.cnn.com/2022/03/31/cars/ikea-furniture-buyback-permanent/index.html.
8. Anelia Milbrandt et al., "Quantification and Evaluation of Plastic Waste in the United States," *Resources, Conservation and Recycling* 183 (2022): 106363.
9. Beyond Plastics, *The Real Truth about the U.S. Plastic Recycling Rate: 2021 U.S. Facts and Figures*, May 2022.

FURTHER READING

Beiras, Ricardo. *Marine Pollution: Sources, Fate and Effects of Pollutants in Coastal Ecosystems*. Amsterdam: Elsevier, 2018.

Bergmann, Melanie, Lars Gutow, and Michael Klages. *Marine Anthropogenic Litter*. Berlin: Springer Nature, 2015.

Cobb, Allison. 2014. *Plastic: An Autobiography*. Aufgabe: 356.

Freinkel, Susan. *Plastic: A Toxic Love Story*. Boston: Houghton Mifflin Harcourt, 2011.

Geyer, Roland. 2021. *The Business of Less: The Role of Companies and Households on a Planet in Peril*. London: Routledge.

Humes, Edward. 2012. *Garbology: Our Dirty Love Affair with Trash*. New York: Avery.

Liboiron, Max. 2021. *Pollution Is Colonialism*. Durham, NC: Duke University Press.

Meikle, Jeffrey. 1995. *American Plastic: A Cultural History*. New Brunswick, NJ: Rutgers University Press.

Weis, Judith S., De Falco, Francesca, and Mariacristina Cocca. 2022. *Polluting Textiles: The Problem with Microfibres*. London: Routledge.

Zweifel, Hans. 2001. *Plastic Additives Handbook*. Hanser Gardner Publications.

INDEX

DR. IMARI WALKER-FRANKLIN is a research scientist at the nonprofit RTI International in RTP, North Carolina.

DR. JENNA JAMBECK is the Georgia Athletic Association Distinguished Professor in Environmental Engineering at the University of Georgia.